AF872785

# The Oak and the Acorn

By

Dean Grissom

ISBN:979-8-234-20352-6
Printed in the USA by 48Hour Books
(www.48HourBooks.com

# DEDICATION

To Gordon and Dottie Luff for their efforts to find a foster family for a broken and damaged young boy. For opening their hearts and home as a temporary safe place.
And for their prayers for me during those years. Thank you.

# DEDICATION

To Gordon and Dottie Luff for their efforts to find a foster family for a broken and damaged young boy. For opening their hearts and home as a temporary safe place.
And for their prayers for me during those years. Thank you.

# PREFACE

Adults sometimes do not understand that children are not these little creatures who are too young to understand certain issues or be impacted by outside factors. The little ones feel and hurt with their limited exposure to life. Their trust of older people is all too often betrayed by individuals around them who are too immature or too incapable of realizing that their own actions can be the making or breaking of the children in their care.

This story is an attempt to show how love can overcome poverty, age and the burdens that life can bring our way. The author's goal was to show how being honest, fair, supportive and loving to those in our care, can create an environment where little lives can flourish and grow into something beautiful.

## CHAPTER ONE

My name is Tommy Meadors. I was born in 1933. Looking back to my early years, I would say that it would have been very hard to have had a better life than mine. The years prior to being in my twenties were spent on a 160-acre farm with my Grandpa Roy and Grandma Ida. Our home was a few miles from Afton, Oklahoma, located in the northeast corner of the state along the Arkansas border.

Grandma was part Irish and Grandpa part Cherokee. He had spent a little time in Tulsa before returning to the farm. They met there. I guess it was one of those loves at first sight things. They were married after dating for only a few weeks.

My early memories are of a home filled with love and conversations that contained words in English, Cherokee, German and a few, hard to understand Irish sayings. We never had much excess, but we always seemed to get by somehow. We had a huge garden, blueberry bushes, blackberry vines, a strawberry patch, two or three milk cows, pigs, chickens, and a horse or two.

I had my own bedroom. Grandpa and Grandma had the other bedroom. Our kitchen always smelled of wonderful and delicious things she made each day. We had a wood stove for heat, but it was also great for slow cooking the evening meals.

Through the years, many additions and improvements had been added to the original house. We had a tub and a sink. Our kitchen was as modern as any around the area.

Electricity had come to our small community a few years back. It was a big change for the folks who lived in town, but unavailable for those of us who lived a few miles away. Life was good in every way. We ate well every day of the week, but Sundays were the best. After getting home from church, which we seldom missed,

Grandma would prepare a feast.

While she was cooking, I did chores, so that after eating I was free to go play. My favorite spot was under a huge oak tree that was growing on the shore of a small lake that occupied the southwest corner of our farm. I would spend all afternoon there.

I do not know what kind of dog he was, but he sure loved to swim. His name was Oganalii which is a Cherokee word meaning "my friend". But I just called him Ogi.

In the summer of 1938, I had been informed that in the fall, I was to be enrolled in the Berry Creek School, a few miles from our farm. My life would be changing with the addition of attending school. Our family could never have anticipated what would result from my starting first grade.

## CHAPTER TWO

September in Oklahoma is still hot and sticky. On the 6th of that month, Grandma got me up early and had me take a bath even though I had taken one the day before while getting ready for church. She had laid out my best pair of jeans and a clean plaid shirt. I only had one pair of boots, but she had cleaned them, and they sat on the floor beside the chair that held my clean clothes.

We had the usual breakfast of eggs, bacon, biscuits and milk. When we were finished, Grandpa told me to meet him outside by our old truck. On the way out, Grandma grabbed me for a hug, smoothed down some stray hairs on my head, and handed me a paper sack lunch. There were a couple of tears rolling down the wrinkles of her weathered face.

I asked her what was wrong, and she just said that I was growing up too fast. Using her apron, she wiped away the tears and kissed me goodbye.

I waited by Grandpa's truck while he told Grandma goodbye. Ogi was sitting at my feet with a stick in his mouth. This was new to him as well. Each morning while I was doing my chores, I would throw a stick now and then for him to fetch. He probably thought I should be throwing the stick he was holding. I told him "tla Ogi" which means "no Ogi". He looked a little hurt and did not seem to understand the situation. Before I could explain it to him, Grandpa came out and we loaded up for the trip to Berry Creek School.

It was only five miles to the school, but it seemed a very long way to a soon to be first grade boy. School started at 8:30 each weekday morning. We were leaving early so Grandpa could sign me up before the school day began.

The road from our ranch to Berry Creek was a gravel

and dirt mixture. When it had not rained for a time, the dirt would boil up into a cloud of dust as a car or truck drove by. Today was one of those times. There was no wind, and we just carried a huge cloud of dust with us as we made our way toward the school.

Grandpa was a quiet man who always measured his words before he spoke. It was normal for me to wait for an answer after I would ask him a question.

As we drove along, I looked at the man who had been my father and best friend my entire life. His huge rough hands gripped the wheel of our truck, and his deep clear eyes watched the road. His skin was weathered and betrayed his Indian heritage. Although he had a full head of hair, it was beginning to turn gray. It was always hard for me to tell if he was in deep thought or just at peace with life.

I finally asked him why Grandma had been crying as we were leaving that morning. I will never forget his delayed as usual answer. "Tommy, mothers and grandmothers love their kids and want to take care of them forever. Grandma loves you more than you will ever know. I think she loves you more than she loves me. Going off to school is the first real change in our lives since you were born. It is hard for her because now she must share you with school time. She is sad to give up that part of the day."

His answer satisfied me even if I did not totally understand the full meaning. Knowing Grandpa, I realized that there would be no further explanation.

We were nearing the school, and I could see it up on a little hill. It was two stories high, square with windows all around the outside. There was a parking lot in front, and a grass covered playground in the rear. Two huge oak trees shaded parts of the rear play area. The front of the building was brick and beside the doors there was a bicycle rack and a large flagpole with an American flag.

Because we were early, there were very few cars in the parking area and only a couple of kids played outside the building. Grandpa parked the truck, and we entered the building through the front door where Grandpa looked at the signs and figured out that we needed to enter the room marked office. The room was occupied by one lady sitting behind a counter doing something on a typewriter. The lady looked up and said, "Well who do we have here?" Grandpa looked at me and said, "Tell her your name." I told her my name was Tommy Meadors, and my grandpa had brought me to the school to go to first grade.

She was very nice and said "Let me get some forms and see if we can get you all settled in for your first day of school. My name is Mrs. Arons, and I am the school secretary.

"Mr. Meadors, why are you enrolling Tommy and not his parents?" Grandpa hesitated before answering. "Would it be ok if you and I talked in private?"

Mrs. Arons thought for a moment and said, "That is not a problem, we can use the principal's office." She and Grandpa disappeared for a few minutes. I just sat there taking in all the things in the school office. There was a large bell attached to one wall. On another was a huge clock. A very large calendar occupied another wall.

Grandpa and Mrs. Arons came back into the room, and she told Grandpa we will just need to have you fill out a couple of additional forms." She opened a file cabinet and retrieved several forms and brought them to the counter. Spreading out the forms, she told Grandpa that she would need my birth certificate, the records that I had received both of my shots, and the legal forms they had discussed in the other room.

Grandpa was quiet for much longer than usual. I was waiting for a reply and did not hear one, so I looked at him. He was obviously unsure of what to say, so he was saying nothing. Mrs. Arons finally spoke up, "Is there a problem Mr. Meadors?"

Grandpa finally spoke and asked Mrs. Arons if he could take the forms home and have Grandma help him fill out everything. She told him it would be fine and that she was sorry there were so many forms, but that it was required that every student have a permanent record of all these things before they could enter school. Grandpa gathered up everything and we headed out of the school to our truck.

There were a lot more children at the school now and they all stared at us as we left. I was not sure why I was unable to go to school. Grandpa drove all the way home without saying a word.

As we drove up to our house, Grandma must have seen the dust and knew Grandpa was coming home. When we stopped and both of us got out, she looked confused. Nothing was said between them, but his look must have been enough communication to let her know there was some kind of problem.

Grandpa looked at me and told me to change out of my good clothes and since I was not going to school that day, to go do my chores. I changed into my everyday work clothes and headed out to the chicken coop. When I left the house, they were sitting at the kitchen table looking at the pile of forms and neither of them was saying anything.

## CHAPTER THREE

I finished my chores and played fetch with Ogi for a while. It was getting about time for lunch, and I had left my sack lunch inside on the kitchen table. So Ogi and I headed up to the house.

They were both still sitting at the table and Grandma was looking at one of the forms. She looked up when I came in and smiled at me. I told her I was getting hungry and came to get the paper sack she had given me. She glanced at the clock and said "Gracious, I had no idea it was so late!" Grandpa said, "Ida, let's have something special and not out of a paper sack."

While she went about getting our lunch together, Grandpa quizzed me about my chores. I assured him I had done all the normal things and then played with Ogi for a while. He told me he and Grandma were unsure of what to write on the forms so after lunch they were driving back to the school and talk to Mrs. Aarons. He told me since all my chores were done, I could take Ogi and go down to the lake while they were gone to the school. That was fine with me.

We ate meatloaf sandwiches made from the leftover dinner the night before. Grandma even opened a jar of her homemade pickles. And she surprised us both with apple pie for dessert. It was the best lunch I could remember.

Ogi and I headed off toward the lake. Before we even got there, I saw the cloud of dust following our truck down the road. Ogi and I swam for a long time. Played fetch for a few minutes and both lay down in the shade and took a nap. It must have been longer than I realized because when I woke up the afternoon sun was moving toward the other side of our fields. I told Ogi we had to head home. When I got close to the house, I could see the truck was back home.

I opened the screen door and walked into the kitchen. Grandma saw me and grabbed a dish towel that was laying on the table. She quickly used it to wipe her eyes. It was obvious she had been crying. Grandpa looked upset as well.

I asked her what was wrong. She just looked at Grandpa and said ,"Roy we need to tell him." I said, "tell me what?" Grandpa looked down at his hands as he rubbed them together and tried to find something to say.

After the usual delay he started "Tommy, I hope you know how much your grandma and I love you. You are the most wonderful gift God has given us. But there are a few things we need to talk about. When we went back to the school we had a long talk with Mrs. Aarons. It seems we have made a few legal mistakes that need to be corrected before you will be allowed to enter first grade."

Grandma started to cry and as I watched those tears run down her face, I thought what could be wrong? Grandpa was not saying anything, and the silence was like a heavy cloud over all of us. I said, "What is wrong Grandpa?" I was about to cry myself and no one was saying anything.

Finally, he said "Tommy, we are not your grandma and grandpa. Everyone in town knew that for many years we had wanted a child of our own. After so many years we just gave up and were convinced that we were just going to be alone. One morning we were awakened by a baby crying out on our porch. Grandma and I jumped up and went to see what was making that noise on the porch. When we opened the door there was a little basket with a tiny baby laying in it.

Grandma picked up the basket and saw a note beside the baby. She handed it to me, and I read "please take care of my baby".

We took you in when you were just a few days old. It seemed to us that God had answered our prayers and given you to us. We named you Tommy and loved you more than you could possibly know.

Everyone around here believes that we are raising a relative's baby. We thought we might get around to telling you someday, but it just never seemed the right time. Now with the school and all these forms, we found out we have not done a lot of things we needed to do.

Nothing will change, but we need to go to town and fill out more papers that make it so us adopting you, is all legal and such. We will need to take you to see a doctor and get you some shots".

When he had finished, Grandma wiped her eyes and tried to speak. With a shaky voice she said "Tommy we never meant to hurt you. From the day we found you, on the porch you have been our little boy. There is not a day that goes by that you do not bring us joy. I cannot imagine life without you. Please understand that we have always done what we felt was best for you."

By the time they had both finished talking, I felt like my world had been turned upside down. All the things in my life, were not real. These were not my grandparents or even my relatives. My real family was who knows where and I was not even Tommy Meadors. It was more than my six-year-old brain could process. It was the worst day of my life.

I ran out the back door and called Ogi and we headed up to the lake. I cried all the way there and Ogi just seemed confused by my actions. I sat in the shade with my arms wrapped around my dog and my face buried in his hair. I cried for so long that I felt there could not possibly be any more tears in my body.

I must have fallen asleep because I woke up to a wet tongue licking my face. Ogi was sitting beside me with sad eyes and a look of concern. I told him the whole story. His eyes never left my face. It seems odd to say but it was as though he understood every word. I thought the only thing to do was go back to the house because it was nearing time for supper.

# CHAPTER FOUR

Supper was more like Sunday dinner than what we normally ate during the week. Grandma was very quiet. In fact, after Grandpa prayed for our dinner, no one said anything for a long time. I just kept my head down and ate my favorite meal, fried chicken, mashed potatoes, and homemade rolls.

When I was almost done eating Grandpa cleared his throat, which meant he was about to say something.

It has been more than 30 years, but I still remember every word he said. "Tommy, Grandma and I have been talking all afternoon. We never understood that this would come up so soon. When you love someone as much as we love you, it just seems that life is so perfect."

" If we had a little boy of our own, we could not love him anymore than we love you. In our family, you are our little boy. I do not know what life would be like without you and I really do not want to know. So, I guess what I am trying to say is that other than having to tell you, about how you came to live here, nothing has changed for us. You are still Tommy Meadors, our Grandson who we love more than life itself. I do not know why you were left in our care, but we are both so thankful you are here with us."

"When I was young and was born on this farm, there were lots of brothers and sisters. My father did not have time to spend with me as I do with you. My mother was so busy taking care of a large family that she never had time to spend with me alone as Grandma does with you. This is our farm, you, me and Grandma. When we die, it will be your farm, Tommy."

"Maybe someday you will take your own little boy down to the lake and show him how to catch a fish or build a raft. I hope you can work through all this and come to understand that we are your family. There are some legal things we all will have to go through, and you will be asked if you want to keep living here with us. Grandma and I sure hope you will want us to be your family."

When he was done speaking, Grandpa folded his napkin and placed it by his plate. He looked over at Grandma who smiled back at him and gave a tiny nod. She was trying very hard to remain in control of her emotions, but there were tears flowing down her face. I just looked at them. First at Grandpa and then over at Grandma. They were my family. I had never known any other. I was raised on this farm. They were my world, and I did feel loved. Grandma had always called Grandpa her oak tree and always called me her acorn.

I had no idea what I was supposed to say or what they expected me to say so I said nothing. I got up from my seat and walked over to Grandma and put my arms around her and just held on. Grandpa came over and hugged us both and we just held on and said nothing. I think we all shed some tears and maybe each of us made a silent commitment to our little family.

The next morning, Grandma woke me up as usual. She told me to go do my chores while she made breakfast. Grandpa was already out milking the cow and feeding the horses when I got to the chicken coop.

When all the morning work was done, Grandma fed us and reminded Grandpa and me that we all had an appointment with a lawyer later that day.

After eating, we all cleaned up and took the old truck to town. I asked about what was going to happen but neither of them was sure of exactly what we were going to be asked to do. We just drove along in silence.

When we got to town, Grandpa drove to an office building on Main Street. It was a two-story brick house that had been converted into a legal office. We were greeted by a lady sitting at a desk as we entered the office. She said her name was Mary and that Mr. Abbott would be with us shortly. She had some papers for Grandpa to look at while we were waiting for Mr. Abbott to see us.

I had never been to an office before and the whole thing was a little overwhelming to me. After a while a door opened and a man a little younger than Grandpa told us to come on in. He introduced himself and shook hands with Grandpa and then me. We each had a chair around his desk, and we sat there for a minute while he shuffled through some papers.

When he spoke, his voice was soft and calming. " Well, I am very happy to meet such a nice family. Tommy, do you know why you are all here?" I just shook my head. He smiled and said "Well it is my job to make sure that we get a lot of forms all filled out properly for our court here in town. These forms will allow a judge to make sure that you are well taken care of and that there are papers to show how he has handled our request to have you stay with your grandpa and grandma. So, the first thing we need to do is write down all the facts."

Grandpa told him the entire story and tried to supply dates when Mr. Abbott asked for them. He took lots of notes and asked lots of questions. Grandma and I just sat there listening. I stared at the wall which was covered by a huge bookcase. I had never seen so many fancy books. The whole place looked like a library that Grandma had taken me to one time.

After what seemed like hours but was actually a few minutes, Mr. Abbott looked at me and said "Now Tommy, I need to ask you a couple of questions. First, do you understand that your grandpa and grandma are going to be asking the court to allow them to legally adopt you and make you their own son? Secondly, I need to ask you if it is your desire to have them adopt you."

I just sat there looking at Mr. Abbott. Grandma finally said "Tommy, you can answer the questions now. Do you want us to adopt you?" I nodded my head. Then he said "Do you understand what that means?"  Again, I nodded.

Mr. Abbott asked if I had any questions.  I shook my head no.  He then proceeded to explain what would happen at the court appearance and that I would be asked the same questions by the judge.  He went on to say that a lady from the court would come to our home and see our house.  She would file some additional papers that the judge would use to help make his decision. He told us that he would file for a court date and let us know when it would be.

He also explained that I would not be able to enroll in school until the matter was settled.  We drove home in our old truck. When we walked into the house, we all sat in the living room sort of worn out from the whole process.

The next morning started as every morning had before this new legal thing interrupted our lives.  I think Ogi was the happiest, because things seemed normal again.  He got to play fetch and go with me to swim in the lake after I finished my chores.

## CHAPTER FIVE

The next several days were uneventful. Life was pretty much the way it had always been. We did not discuss the court appearance that would be coming soon. My days were filled with chores, and trips to the lake with Ogi. Both Grandma and Grandpa seemed happy but a little reserved. I was glad not to be going to school.

Our garden was not producing as much because the season was changing. And the weeds had stopped growing which made my chores less.

Our little farm consisted of a few acres of corn which was used to feed our pigs and chickens. There were two pastures where our cows and horses could graze. And there was a little spring which produced the drinking water for our house and for our animals.

On Saturdays, one of my chores was to haul water buckets from the spring to three large oak whiskey barrels that Grandpa had placed on a wooden stand next to our house. These barrels held water from the spring, and they were attached to a pipe that allowed water to flow from each one, into our kitchen.

We did not have electricity or a telephone because we lived too far from town. Our light at night was from several kerosene lanterns and candles. We also had an outhouse for daily needs.

Grandpa would go to town once a week to get the mail from the post office, pig feed from the mill, and flour, sugar, cornmeal for Grandma from the general store and a newspaper. The only time that this routine changed was when we had extra pigs to sell or there was a holiday coming.

Our food was provided by our garden, our chickens, and our pigs. Occasionally, we would raise a steer and have beef for the next few months. We ate well and I never realized until many years later that we were considered by most standards to be poor people.

There were some farmers in the area who had purchased tractors after the war. They used these to cultivate much larger farms than we could ever have because we still used horses to work our land.

Grandpa was watching our corn, and he felt that we would be able to start harvest in the next few weeks. It was always hard to decide when to pick the corn. It needed to be mature and dry, but the fall rains could destroy an entire crop. Timing was everything. I secretly hoped that we would start harvest soon and that way I could avoid going to school.

The days rolled by and we had not heard anything from the lawyer or the court. One day a neighbor was returning from town and stopped by with a letter from the post office. It was from the lady the court had appointed. She was coming to visit us on the following Monday.

We all spent the next few days making our little home look warm and welcoming. On Sunday after we returned from church, Grandma made a little cake for the next day.

Monday finally came and the lady from the court showed up around mid-morning. I was wearing my Sunday clothes and Grandma had me take a bath that morning.

Grandpa answered the door and welcomed the lady inside. She introduced herself as Mary Owens. We all sat down, and Grandma offered Miss Owens a glass of sweet tea and some of the cake she had made.

After some polite conversation, Mary pulled out some papers from the briefcase she was carrying when she came in the house. She asked Grandpa and Grandma a lot of questions about how long they had owned the farm and how long they had been married. Then she asked to see the rooms in the house. During all this activity she was writing notes on her papers.

When she finished with all of that, she started asking me questions about the chores I was asked to do and what I did in my spare time. It seemed important to her that I was happy living there.

I assured her that I was treated very well and would not want to live anywhere but here in this home. When she finished asking all the questions, we were all waiting for what was coming next.

After she had put all her papers away, Mary told us that the next day she would file a report with the court, and it would be delivered to the judge who would eventually have an adoption hearing for them. Grandpa asked when that might be, and she told him she was sure it would be within two weeks.

Mary thanked us and said she had enjoyed her visit. We all wanted her to give us an opinion or some impression she had from the visit. But it was not to be.

When Mary was gone, we sat there for a while. Grandma was the first to speak. She said, "Well I think that went pretty well and I feel positive about the whole thing". Neither Grandpa nor I said anything. I changed into my normal clothes and said I was going to take Ogi to the lake. Grandpa said that it would be fine.

Grandma was in the kitchen preparing for dinner and Grandpa was sitting in the same chair he had been in when Mary left.

Ogi and I ran all the way to the lake. When we arrived, I sat down and pulled him into my arms and proceeded to tell him about the whole day with Mary Owens. I do not think he was very interested because he kept eyeing a stick that was nearby. I gave in and released him so he could grab the stick and bring it to me. We played fetch for over an hour.

I spent most of that hour trying to make sense of school, finding out I was an orphan, not knowing if I would continue to live in my home or be given to some stranger. It was far too much for me to understand. I could not figure out how people who did not even know me could decide all these things about my life.

Ogi was finally tiring out, and I was getting nowhere trying to understand all these adult things, so we headed for home. All that I knew for sure was that corn harvest was about two weeks away and the court hearing was about the same. If we started harvest maybe we would not need to go to the hearing. And without the hearing, I would not be attending school.

When Ogi and I got home, Grandma told me to wash up and sit with Grandpa until dinner was ready. He was still in that same chair. I sat down and waited for what I do not know. He finally glanced up at me and said, "So what did you think about Mary?" I told him that she seemed nice, but I was still confused about the whole day. He just listened as I told him that I wanted to stay with them and did not understand how strangers had anything to do with what we wanted.

He gave a long sigh and said, " Tommy, there are rules and laws that were put into effect, to help guide people along life's way. Most of them are good rules. But sometimes there are circumstances where the rules do not exactly fit the situation. It does not mean the rules are bad. But sometimes the enforcement of those rules can have unpleasant results."

"In our case, we all love each other, and we have a happy life together. In some families there are problems, and the rules say that to protect the children, the judge can take them away from a bad family and give them to some family that will love them. So, the judge needs to look at each case and decide if the orphan is in a good home or a bad home."

"Mary came here to see if you wanted to stay and be our grandson. Her job is to report to the judge that you are happy here and that we love you with our whole heart. Her report helps the judge understand our family. And that helps him decide if it will be good or will not be a good thing to let you stay with us. I believe the judge is a good man and will see that we are a good and loving family. I believe that God wants us together. Do not be afraid Tommy. It will all work out."

I listened carefully to all he had to say. It did not make me feel at peace with my future, but it did make me feel a little better.

Grandma appeared from her work in the kitchen, and we were summoned to the table. Grandpa told me that tonight was special, and I should go and see if Ogi wanted to come in and visit while we ate our dinner.

I ran to the back door and called him. He was excited as this was unusual. Grandma put all the food on the table and gave each of us a plate. She even had one for Ogi. It was my favorite meal. Fried chicken, mashed potatoes and gravy with fresh rolls.

Before we ate, Grandpa prayed and thanked God for our meal and asked Him to make all the legal things work out so our little family could stay together. Everyone seemed to feel a bit better, and we did not discuss the future anymore during dinner.

Ogi had mashed potatoes and gravy, which disappeared long before anything on our plates. Grandma surprised us with blackberry cobbler for dessert. For a moment in time, all was well.

## CHAPTER SIX

The next few days drifted by with very little change in our daily routine. Grandpa had gone to see our closest neighbor Mr. Amos. He had a large farm that formed a border to one side of our property. Mr. Amos had purchased a corn harvester last year and Grandpa had taken me to watch it work. It was amazing. What would have taken us an entire day on our farm, that machine accomplished in less than one hour.

When Grandpa returned from the visit next door, he told me that we were going to hire Mr. Amos to use his machine to harvest our corn. I was very disappointed because I enjoyed working in the field and now did not have any excuse to miss school.

Grandpa said he was going to town in the morning and asked if I would like to go along. That made me happy. A trip to town often included a few pennies which I could spend at the general store buying candy. He told me to do my chores in the morning, and we would leave after he was done with his work.

Grandma had prepared a list of things for him to pick up while we were in town. On the way to town, Grandpa said our first stop would be the post office to see if there was anything from the Judge. After that we would make all the normal stops and one of those would include a few pennies for candy.

It was almost noon before we were able to make it to the post office. Ellen, the postmaster, saw us walk in and she went over to a stack of mail and pulled out one piece and handed it to Grandpa. She told him it had come in the day before, so she had put it aside for him. Grandpa thanked her and we walked out to the truck.

He turned the letter over in his hands just looking at it.

"Tommy", he said, why don't we wait and open this with Grandma when we get home?" I think it was more of a statement rather than a question.

We did our usual stops feed store, general store, newspaper office, and kerosene. On the way home neither of us spoke about the letter. I ate some of my candy and sat quietly on my side of the truck.

When we got home, Grandpa and I unloaded all our supplies before taking the letter and groceries in the house. Grandma was sitting in her favorite chair doing some sewing. Grandpa told me to carry in the rest of Grandma's supplies, and he sat down in the chair beside her and handed her the letter.

Neither of them spoke they just looked at each other and the unopened letter. I finished bringing in all the packages from the general store and walked into our parlor and sat down. It was just quiet for a long time before Grandpa told Grandma to open the letter.

She slowly tore the envelope open and removed the one-page document from inside. We both watched her unfold it and start to read. Her expression did not reveal what she might be feeling as she read the entire letter before speaking.

"It says that we are all to appear before the judge next Monday morning at ten o'clock" she said. And then she handed the letter to Grandpa who also read the entire page.

I wanted to know every word that was written in the letter, and I finally said, "is that all that it says?" Grandpa said that there was a lot of legal talk, but the main message was just for us to appear in court.

This being Saturday, we would all attend church the next day and then go to the court on Monday. I wanted it all to be over. Monday seemed so far away. How could we wait so long to hear the judge decide. I looked at both their faces, and I think they were feeling the same emotions that I was feeling.

Grandpa looked at me and said, "Tommy get Ogi and go play by the lake. Maybe take your fishing pole and see if you can catch us a fish. Be home by dark and don't worry about your chores, I will take care of them today". Grandma told me to cut a piece of the blackberry cobbler for a snack. They did not have to tell me twice. In just a matter of minutes, I had my snack, my fishing pole, and Ogi and I were headed for the lake.

We got to the lake, and I dug up two worms for bait. I put one on my hook and attached a bobber to my line. There was a favorite spot of mine down the bank a bit, and I cast out my line and watched it settle very close to where I had been able to catch fish on many other trips.

After that, I ate part of my snack and then found a stick to play fetch with Ogi. We played for a long time and between throws I would look at my bobber to see if there was any activity. Ogi did not seem to want to play anymore, the fish were not biting, and I was ready for a nap. Ogi and I curled up under our favorite tree and I was soon asleep.

It was sometime later when Ogi started barking and woke me up. He ran over toward my fishing pole. I could see my bobber racing across the water surface. There was a fish on my line for sure. After grabbing my pole, I started reeling as fast as possible.

It was a good size fish, and it jumped out of the water more than once. I never carried a net, so it was just me against that fish. He finally tired and I pulled him in with the fishing line. Just as I reached down to grab him, he took one last flop and came off the hook and slid back into the water. It was disappointing because I really wanted to show that fish to Grandpa. But secretly I was happy for the fish. Maybe the next time I hooked him, I would feel differently. I told Ogi we better be heading home. It had been a wonderful afternoon.

Ogi and I arrived home in time for dinner. Grandpa asked me how the fishing had been, which gave me the opportunity to tell him about the one that got away. He just laughed and said that I would get him the next time.

There was not much talk during dinner and afterwards Grandma reminded me that church was tomorrow and that I needed to take a bath. After getting clean, I just went to bed, tired from the emotion of the day and my trip with Ogi.

Sunday was no different than any other Sunday. Church, fried chicken, chores, leftovers for dinner, and going to bed early. We did not talk about the next day beyond setting a time to leave the house for the trip to town.

I had a hard time going to sleep. The meeting with the judge was the biggest event in my short life of six years. Eventually, sleep overcame the worries.

We ate in silence and loaded up for town. When Grandma got me up on Monday morning, she had already prepared eggs, bacon, and biscuits.

We drove to the courthouse in silence. It was not a happy time in our little family. The short walk from the truck to the door of the court seemed like a long journey. My feet felt heavy, and my legs seemed weak.

When we entered the building, Grandpa asked someone where we were supposed to go to see the judge. A nice lady led us over to a door and after opening it, showed us where we could sit. It was a little before ten and we were the only folks in the room.

The quiet in that room did nothing to calm my fears. What seemed like a long time was just a few minutes before the door opened and an older man took a seat at a desk in the front of the room.

He looked at a pile of papers then glanced at the clock on the wall before saying "My name is Judge Winslow, and we are here today to render the court's decision on the matter of the adoption of Tommy Meadors."

He went on to say that before he would make the final statements that he wanted to ask a few questions. He told us he would start with questions for Grandma and Grandpa and then ask me questions after they had finished.

He looked at Grandma and said that he would like to hear from her first. "Mrs. Meadors, at your age do you feel that you can be a mother to this child?" Grandma thought for a minute and then said, "I was never able to have children of my own. When Tommy came along, I thought that God had answered my prayers and given me a child of my own. I could not possibly love him anymore if he were my own blood. He is my life, and I cannot imagine one day without him." The judge thanked her and then addressed Grandpa.

"Mr. Meadors do you think you can be a good father to this child at your age?" Grandpa stood up and said, "Judge, I look at Tommy each day and thank God that he sent this little boy to fill that empty place in our hearts. Our family would not be complete without him. I believe he knows how how much we love him.

Grandpa sat down and the judge looked at me for a while and then said, "Tommy please tell me why you want these folks to adopt you". I stood up as Grandpa had done and started to speak. The words were stuck in my throat. The judge told me it was ok and to just take my time.

When I could finally get the words out, I said, "Sir, all my life I believed that these were my grandma and grandpa. I never wanted anyone else. They love me and I love them. I do not want to live anywhere else. I have my dog Ogi, and I help Grandpa with the chores. Grandma knows my favorite meal and makes it for me once each week. I can go fishing at the lake and once a week Grandpa gives me pennies for candy at the general store. I do not know how I could be any happier. Please do not make me go live with strangers". Then I sat down.

The judge looked down at his papers for a long time. When he finally spoke, he said the words that would forever be burned in my memory. "Mr. and Mrs. Meadors it is the findings of this court that the minor Tommy Meadors has a good home filled with love and that he will be allowed to be legally adopted by you. Congratulations to you all. Tommy you are a very lucky boy to have such loving grandparents. Good luck to you all".

He closed the file and smiled at us and said, "Case closed, have a good life. We will prepare some documents for you to sign the next time you are in town".

I was not sure what to do. Grandma was crying and Grandpa held her in his arms and reached for me and we just all stood there holding each other.

After a minute or two Grandpa said, "I think we need to go to the store an get an ice cream cone". And so, we did.

The ride home was quiet but for a different reason. I think we were all just overcome with relief. We were a family again and forever. Once again, Grandma had her oak and acorn.

## CHAPTER SEVEN

When we got home, Grandpa told me to change my clothes and take Ogi to the lake. He said he and Grandma had to make some plans about school and shots.

Grandma made me a snack, and I ran out the door calling for Ogi. On the way to the lake, I told him everything that I could remember about the day in court. He did not seem very interested. I think his mind was on swimming and fetching sticks.

We enjoyed the afternoon and got home by dinner time. While Grandma was setting the table, Grandpa told me he was going to town in the morning to see the doctor and the school principal. He told me to do my chores in the morning and then help Grandma with whatever she might need.

By the time that I was out of bed the next morning, Grandpa was already gone to town. I did my chores and Grandma wanted me to count how many hens we had and how many young pigs. She did this a few times a year because we were able to sell our excess critters to help replenish her cookie jar fund. After finishing all those chores, I was not interested in going to the lake, so I just sat outside under our tree and fell asleep. The noise of Grandpa's truck coming down our road woke me.

He told me to come inside that he had some news. When all sat down, and he related everything he had found out at the school and the doctor's office. The school told him that all he needed to do to finish my enrollment was to bring a copy of the adoption paper and a certificate from the doctor showing that I had received my two shots. He also said that the court would have the papers ready the next day, so he and Grandma would go to town and sign those and then take me to see the doctor.

All this news was good, but I did not want to hear about going to school. Grandpa went back out to his truck and brought in a newspaper. I could not read what it said but he read it for us. The paper had big letters on the front which Grandpa said, told the story that England was going to war against Germany. Since we did not have electricity, telephone or a radio, we were unaware of any of the news printed in the paper.

When Grandpa finished reading the article, there was silence. I did not really understand what any of that news meant to our family. It was obvious that it did mean something to Grandma and Grandpa.

Both of my grandparents had lived through the war in Europe. They knew what effects a war had on the people of a country. Grandpa brought up several issues that could make an impact on the United States even if we were not involved in another war.

I was more worried about getting shots and going to school. I did not really know where Britian or Germany was located. We made our plans for the following day, had our dinner and went to bed.

The next morning was all just the usual until we were dressed in good clothes and heading to town. There was lots of uncertainty in our little family. Our first stop was the doctor's office. Grandpa checked us in with the receptionist.

We did not have to wait long before a nurse came out and called my name. She greeted me and told me her name was nurse Annie. She explained that it would not take long and started to take me into the back room. I hesitated long enough for her to realize that I was afraid. Annie asked me if I would like my grandma to come with me. I shook my head yes, and we were both ushered into a small room.

Laid out on a metal tray was a long needle and two

bottles of some substance. We had just settled into our chairs when the doctor came in and introduced himself as Dr. Frank. Grandma knew him and just smiled.

He stuck out his hand and shook mine. He said, "This will not take long and will not hurt." While he was getting the needle ready, he told Annie to fill out the forms and get the certificate ready for the school.

When he was ready, he rolled up my sleeve and rubbed something on my skin. And asked me about Ogi. Before I could answer, he said, "Ok, one more." The second one was far less scary for me.

Annie had some papers ready for him to sign and when he was done, he said, "Ok Mr. Tommy, you are all set. Good luck in school". He told Grandma which paper to give to the school and thanked her for coming in with me.

We left the doctor and drove to the school. We all went to the office, and Grandpa presented the health certificate to Miss Arons, along with a copy of the adoption papers. She looked them over and told us that everything looked in order and I could start school that day or wait until the next day and start with all the other children at 8:30.

I did not wait for anyone else to speak, I told her I wanted to wait until the next morning. She told us that it would be fine and to come to the office in the morning and she would take me to my class and introduce me to the teacher and the other students.

Grandpa thanked her and we left for home. No one said much on the way home. It had been a trying experience for all of us the last week or so. I think we were all tired and just needed a day with no new surprises.

Grandpa and I did the chores, when we got home, while Grandma made some lunch. Ogi wanted to play fetch, but I was not in the mood. I could not stop thinking about having to start school in the morning.

I did not know any of the kids or my teacher. We lived so far from town that there had never been any opportunity to meet anyone my age. The closest farms to ours were owned by families who had children who were grown and had already left home. I was the only six-year-old who lived in our area.

I wondered if the other kids would like me. I hoped the teacher would be happy to have me in her class. I did not know what a student in first grade would learn. Too many things to think about, so I put my thoughts aside and wondered what Grandma was making for lunch. I would find out soon enough for she was calling for us to come in the house.

The remainder of the day went quickly. I asked Grandma what I was going to learn in first grade. She told me math, reading, writing and some history.

I already knew some math because Grandpa had me count the chickens and the baby pigs. I was able to read the labels on the pig food bags and the chicken feed sacks. Maybe school was not going to be hard for me.

Grandma had my clothes all laid out for in the morning. She said that I should go to bed early because tomorrow was going to be a big day.

Going to bed early was ok, but it did not mean that I could go to sleep early. This was one of those times. It was very late before my eyes were too heavy to stay open. Despite the many things I was thinking about, sleep finally overcame my tired body.

## CHAPTER EIGHT

Grandma got me up and already had breakfast ready. When I sat down at the table, I could see a lunch sack she had packed for my first day of school. We ate quickly and Grandpa brought the truck around front, and we were off to town. Grandpa drove to the school without saying anything. When he was thinking it was my job to remain quiet.

When we arrived at the school, Grandpa headed for the office with me in tow. Mrs. Arons rose to greet us and said, "Well Tommy, are you ready to meet your teacher?" I just nodded yes. Grandpa hugged me and told me he would pick me up after school.

Mrs. Arons grabbed my hand and led me down the hall to a classroom marked 1&2. She knocked on the door before opening it. There was a young tall lady standing behind a desk. We walked over to her and Mrs. Arons said "Tommy Meadors, this is your new teacher Miss. Walters." Then she left us and headed out of the room.

Miss Walters turned me around to face the room. She pointed to a desk in the second row and told me that it was going to my desk. She explained that it was not taken because the student who had been sitting there had moved away recently. She showed me where to put my lunch and a coat if I had one. By now a few other students had drifted in and I was feeling a bit awkward. I sat at my desk and just observed what was happening.

A few minutes later a bell started ringing and lots of kids entered the room and took their seats. Miss Walters was writing something on the blackboard as the bell stopped ringing.

When she had finished, she looked out at the class full of little faces and said "Good morning, everyone. We have a special treat today. Please stand up, Tommy. This is our new student, Tommy Meadors. He lives out of town on a farm with his grandpa and grandma."

Then she looked at me and said, "Tommy why don't you tell the class if you have any pets and what your favorite food is and anything else you think they might like to know about you."

Time stood still for me, and I could not get any words to come out. There were 15 pairs of eyes, looking at me, and waiting for me to speak. Finally, I was able to say, "My dog's name is Ogi, and I like fried chicken. My grandpa is part Cherokee Indian. We have horses, cows, pigs and chickens." I said no more, and Miss. Walters stepped in to help me. "Tommy, Ogi is an odd name for your dog, does it mean something", she asked.

"Well, his real name is Oganalii, which means my friend, in Cherokee. But I just shortened it to Ogi", I said quietly. Miss. Walters addressed the class and asked if anyone would like to ask me a question.

One girl asked if I was an Indian. Miss Walters interrupted and answered the question by saying, "lots of people in our area had ancestors who were part Indian and that made those people more interesting."

There were a few more questions before Miss Walters said that we needed to move on with the day's lessons. I was very happy that she took charge because I was not comfortable with all the things people wanted to know about me.

There was never any time in my life where I had been in a room with fifteen other children. I had no idea of what to say to them or how to respond to them looking at me. So far, my first day in school was not a fun experience. I had only been there a short time, and I was ready to go home.

The remainder of the morning was a blur to me. I do remember Miss Walters writing some things on the black board and her reading to us from a book. But mostly, I remembered the other children whispering and pointing at me.

The clock on the wall eventually showed that it was noon. Grandpa had taught me how to tell time, and my stomach was also letting me know it was midday. Miss Walters told us to collect our lunches and line up to go outside, where I learned we would eat when the weather allowed.

There were several tables outside and as the class was dismissed for lunch, the kids ran to what apparently was their usual spot at certain tables. I was left standing with my paper sack in my hand.

A small red-haired girl finally approached me and told me I was welcome to eat at her table. She told me her name was Maggie Scott. I sat down at her table and opened my lunch. Every child at the table stopped eating and watched as I took out a piece of leftover fried chicken and a biscuit.

Maggie asked me why I did not have a sandwich and an apple. I was completely unprepared for a question like that. It seemed that my chicken and biscuit had somehow violated an unwritten rule. It was apparently the normally accepted thing, to bring sandwiches to school.

The looks of the other students and my humble lunch made me lose my appetite. After taking a few bites, I put the rest of the lunch back in my sack and told Maggie I needed to go to the outhouse. Everyone must have thought that was funny as they began to laugh. Maggie said "Tommy we do not have an outhouse. There are bathrooms inside the school. I will show you where they are located."

My humiliation was complete. School was a terrible place. Maggie led me down the hall to a door with a little boy painted in the middle. She told me to go into the room, and she would wait for me.

I entered the door and looked around. There were little doors and behind each of them was a white place to sit. But there was also a whole row of some white things attached to the wall. I had no idea what those were, so I went into the little room behind one of the doors. When I finished and looked around there were some sinks attached to another wall, and I was able to wash my hands like we did at home.

I walked outside and Maggie was waiting for me. "Well, how did you like the bathroom, Tommy?" Maggie said with a smile. The way she asked was calming and reassuring rather than making fun of me. So, I told her the truth, it was a new experience for me. She just smiled and took my hand the way Grandma would and took me back outside.

On the way she explained that we would have a thirty-minute play time that was called recess. She said the boys usually played baseball and the girls played tag or kickball. I told her that I did not know how to play those games. She asked me what games that I usually played, and I told her that I had never played any games, and that there were no other children who lived near our farm.

She seemed totally confused by my story, but not in a bad way. Maggie thought it might be better if I did not tell the other children that I had never played games. It would be accurate to say that Maggie became my first and then best friend in the coming years.

She sat with me until recess was over and walked back to the classroom with me. The rest of my first day was uneventful. I had learned a few names of other children by listening to Miss Walters call on them during class.

The afternoon went more slowly than the morning, but it was all new to me, so I was able to keep focused. The bell finally rang, and we were dismissed. On my way out, Miss Walters said "I hope you enjoyed your first day Tommy". I just nodded and hurried out to meet Grandpa. He was waiting in our truck and when I got in, he wanted to know all about my first day.

When we arrived home, Grandma was standing on the porch with a smile on her face. She grabbed me and hugged me for a long time. Life was good. Back home with those who loved me. Grandpa told me to do my chores and at dinner we could hear about my first day of school.

Ogi followed me around the barn as though I had been gone for days. He did not want to leave my side. When all the chores were done, I told him to bring me a stick. He was gone quite a while, must have been looking for the perfect piece of wood. We played until Grandma called us for dinner.

We had pork chops and green beans. It was delicious. When we were done eating, Grandpa said "Ok Tommy, how was your first day of school?"

I was unsure of exactly where to start so it came out all mixed up. The adventure of the bathroom was first, followed by lunch with Maggie and the fact that I did not know how to play the games at recess. Grandma asked about Miss Walters. I told her that she was very nice, so it was easy for me to talk about her. It was difficult to say whether I had learned anything the first day, but it was easy to say that school was not very much fun.

Grandpa told us that the newspaper he picked up in town today had said that the war in Europe was a very bad thing and could create long term problems for our country. We cleaned up after dinner and Grandma said that I should take a bath before going to bed.

Taking a bath in our home was a lot of work. We had to fill

buckets of water from the sink, heat the water on our wood stove and pour the heated water into our tub. So, we did not bathe every day. Grandpa told me that we would leave at the same time in the morning. I went to bed wondering what new things would happen at school the next day.

The next morning while we were having breakfast Grandpa said that I would need to get up earlier so that there would be time to do my morning chores before going to school each day. We left for town and Grandpa dropped me off at the school. He said he had a few errands to run before heading home.

This being the second day at school, I knew about the morning bell which started the day, the bathroom, lunch time, recess, the bell ending the day, and my new friend Maggie. But I was unprepared for some new things which I was about to experience.

The second day was better already because I did not have to talk about myself or stand in front of the class. Miss Walters had a list of words and sounds that she had drawn on the blackboard. We were told to open our desks and take out the tablet that each student had been given. Inside there was a pencil and she asked us to try and copy the first word that she had written on the board.

That is how our morning went, saying words and trying to copy the letters on our tablet. I recognized many of the letters from words on our feed sacks in our barn.

By lunch time we had covered several words and even a couple of phrases. Miss Walters dismissed us and after picking up our lunches we headed outside. That is when the troubles started.

Maggie and I headed out to the table where we would eat. We had just started to spread out our lunch when an older boy approached our table. He said, "Would you like to play baseball with us during recess?" I thought he must be a nice person to want to include me in a game of baseball. I told him that I appreciated the offer but that I did not know how to play baseball.

Maggie reached for my hand under the table and squeezed it. I looked at her and she shook her head ever so slightly in a "no" fashion. She looked at the boy and said, "Hank, he does not know how to play baseball. He lives on a farm outside of town. No one out that way plays baseball. They do farm things."

I guess he was not satisfied because he started laughing and said, "Are you a little sissy? I heard you don't even have a family. You are adopted and live with two old people."

Maggie said, "You leave him alone Hank Evans. Go play with you buddies." Hank said, "Shut up you little pipsqueak. I will do as I please. Why do you want to eat with a little sissy boy anyway?" Maggie informed him that she would rather spend time with a boy who had manners than a jerk like he was.

He said nothing more and just walked away laughing. Maggie told me he and his two older brothers were the bullies in the town. Their father was as mean as they were and nobody in town like any of them. She told me that I would need to stand up to Hank at some point or he would make every day miserable for me.

I did not know it at that time, but the entire Evans family would play an important part in my future life. It would not be a pleasant experience.

I spent recess with Maggie, and she began educating me on which students were nice, and which were not. I did not understand why she had decided to be my friend and protector, but I sure like her.

The day was warm, and I could barely keep from falling

asleep. The bell rang for the end of the day, and I was sure ready to go home.

Grandpa picked me up after school and as we were driving home, I asked him about the Evans family. He gave me one of his long looks and asked me why I wanted to know about them. I told him how Hank had treated me and what Maggie said to him. I told him that Hank called me a sissy boy because I did not know how to play baseball.

Grandpa was silent for a long time. He finally cleared his throat and said," Ed Evans was born in town. His father was the town drunk. Their family was always poor because the father was always drunk, and no one wanted to hire him. Their mother was a nice person who tried to raise the three boys to be good people, but the father was hard on all of them. She finally reached a point where she could no longer live with him and one day when he sobered up, she was gone. No one ever knew where she went. Just up and left him and the three boys."

"The boys grew a little older and started stealing and running wild with no father to keep them in line. The oldest boy Matthew, ended up in jail at fourteen years old. He stole a car in town and ended up wrecking it on a speeding trip through the county."

"The other two boys are in the school with you. They both flunked more than one grade and that is why they are the oldest kids in their class. They are bullies and each of them is as mean as an inada (snake)." He warned me not to trust them and to never go anywhere with them. It was his opinion that they wanted to provoke a fight with me and that was why Hank called me a "sissy boy".

By the time Grandpa had finished talking, we arrived home. He told me to change clothes and get my chores done and then come and find him.

I found Ogi and after changing my clothes we headed to the barn. I finished my chores and threw a stick to Ogi for a few minutes. Then I went to find Grandpa.

He was sitting in our parlor. I told him I was all done with my work. He got up and we went out to the barn. In a corner of the barn was a large stack of items that we no longer used. Things like broken oil lamps, bits and pieces of old iron parts, some old bridles from our horses, and a few worn out tires. Grandpa started moving stacks of these things to another area.

After a few minutes his efforts revealed a canvas tarp. As he removed the tarp, I could see an old trunk. It was covered with dust and a few spider webs. He brushed them aside and opened the trunk. Inside were lots of papers, some little wooden boxes, and a second tarp wrapped around some items. On the tarp, written in paint, were the words siga tsagila (baseball). When Grandpa unrolled the tarp, there was a baseball, a baseball glove, a baseball bat, and a baseball cap.

He took each of the items out of the trunk and laid them on the ground. Then he carefully rearranged the contents of the trunk and closed the lid. We put the tarp back over the trunk and then he spoke. "Tommy, these are the things used to play baseball. They were mine when I was young. I am going to help you understand the game and how to use each of these items. We will start after dinner tonight".

And we did. That night and every night after that for as long as I can remember, Grandpa worked with me on throwing the ball, catching the ball, and hitting the ball. During every lesson, he explained the rules of baseball.

I would eventually learn that Grandpa had been a very good baseball player. He had even played on a Cherokee baseball team when he was out of school. I learned that he was forced to give up his dream of becoming a professional baseball player because his father had become very ill, and he was needed on the farm to keep it running and make a living for his family.

During our nightly baseball time, I learned so much more about this man that I called Grandpa. He was normally quiet and had very little to say. These times together seemed to bring out a part of him that was seldom seen. When he was relating his stories each evening, he seemed younger and it was my pleasure to experience his history, that he was revisiting with me.

My baseball skills were improving and Grandpa told me that we would keep our little secret to ourselves for a while longer. He did not want me to play baseball at school until I was able to really demonstrate my skills.

On the weekends when my chores were finished, I no longer threw sticks for Ogi to fetch. I threw my baseball which he was more than happy to retrieve for me.

Grandpa nailed an old tire to the side of barn and had me throw the ball hundreds of times, trying to have the ball hit inside the tire. It was my own personal target. As I would master that activity, Grandpa would move me farther away from the barn. The new distance was more difficult. He was trying to improve my ability to throw the ball accurately at longer distances.

My lessons continued for the entire first grade. My friendship with Maggie continued that entire school year. She was always there for me. I managed to stay away from much conflict with the Evans brothers. My secret baseball lessons remained between me and Grandpa.

By the end of the school year, I had learned to read and write many words. We learned how to add and subtract. I made a few more friends, but none were as nice as Maggie.

## CHAPTER NINE

Summer was hot as usual. We had been lucky the seasonal rains had come on time. Our corn was looking good this year. Mr. Amos not only made a deal with Grandpa to harvest this year's corn crop, he also had an agreement to use his machines to plant the corn that was now growing in our field.

Grandpa had a good relationship with Mr. Amos, and they had agreed that Mr. Amos was to receive a share of the crop rather than Grandpa having to worry about coming up with cash.

Summer was fun because there was no school. However, there were additional chores. When my normal chores were done, I would spend much of the day hoeing weeds in the corn fields. We needed to do this until the corn was as tall as me. Once it was that tall, the weeds sort of gave up their fight to be in the field and just quit growing.

When my time in the corn field was finished each day, Ogi and I would go down for a nice cool swim in the pond. In early July, Grandpa told me the corn was tall enough and I could work on my baseball and play with Ogi after my morning chores.

I had grown a little taller that year. My boots were getting tight, and my working clothes were approaching being too small.

Grandma knew a lady from the church who was friends with a family who had quite a few clothes that their two sons had outgrown. Those clothes had been offered to our family.

I knew who the kids were because I had seen them at school. Both the boys were older than me, but I had never really met them. Grandma announced that when we went to town the next Saturday, that we were going to stop by their home and pick up my new clothes.

That summer brought with it a new part of my baseball training. Grandpa began teaching me how to hit the ball with a baseball bat. It was our time to be together. Grandpa would pitch the ball to me and if I was able to hit it, Ogi would fetch it and drop it at Grandpa's feet. Grandpa seemed to be so happy when we were playing our little baseball game. He just seemed more alive and even younger.

As my hitting skills became better, Grandpa would throw the ball harder, making it more difficult for me to hit. I never complained because I did not want to disappoint him. By the end of summer, I was improving in hitting, throwing, and catching the ball. What made me the happiest, was that I could see that Grandpa was happy. The memory of those days of playing ball with him would stay with me my entire life.

It was never just about baseball. None of life's burdens could steal away those moments we were together. Our little family had no idea of the life changing events the next year would bring.

We were content with life the way it was for us. We would never again have our little peaceful world where the three of us were so accustomed to living.

School started and not much changed. Because the first and second grade were in the same room, we had our same teacher, same classmates with the addition of the new first graders, the same recess time, and some of the same math and reading studies.

I suppose the main differences were, we were all a little taller and the teacher expected more from us as we learned more words in our reading, spelling and writing lessons.

My wonderful friend Maggie had also grown a little. Her hair was longer, and it seemed hard to believe, but I think she had even more freckles this year. We spent our lunches together sharing all that had happened during the summer break.

I was not asked to play baseball at recess, so I kept my practices with Grandpa a secret. Hank Evans had been passed to a higher grade at the end of last year, so he was playing with the bigger kids and so far, he had not talked to me. Life at school was ok. My reading skills were improving.

One day Grandpa picked me up after school. Instead of going home we ended up stopping to buy a newspaper. This was unusual because Grandpa had specific things he did on the days, we were in town. I asked him about it, and he said that he had heard there was going to be a meeting about installing electric lines into some of our rural areas near town. There was supposed to be a map in the newspaper that showed the new proposed electric lines.

When we arrived at our house, he headed straight in and called to Grandma. He spread the newspaper out on the kitchen table and turned the pages until he saw a map. Together they poured over the map for a few minutes. Grandpa pointed to an area on the map and said, "Well, I cannot believe it, but we are going to get electricity and there are rumors that we will have a telephone line as well."

He read the whole article out loud. There was a meeting the following Friday and everyone who was on the roads where the new wires would be installed was asked to be present. Grandpa announced that we were all going to the meeting.

The days dragged by leading up to the big meeting. Our little family had talked every night about how electricity and phone service would change our lives. Grandma wanted lights in the kitchen. Grandpa wanted to buy a radio and have it in the living room so he could listen to ball games. I wanted a light in the barn so I could see to do my chores. Looking back years later, it seemed our dreams were so basic.

The big day finally came, and we walked into the town hall. It was very crowded, and folks were greeting each other as they found a seat.

At the appointed time a man dressed in a suit walked to the front of the room and stood behind a desk. Behind him on the wall was a very large map showing our town and the roads out into the farms. Several of these roads had red lines drawn on them.

He cleared his throat and said, "My name is Norman Petters, I am the vice president of Rural Electric Company. We are here tonight to discuss bringing power and phone service to the farms within a seven-mile radius of this building." There were some excited murmurs from the people in the room.

Mr. Petters went on to say that the construction cost of the project would be paid by his company, but there would be a small fee that each farm would be required to pay before we would be allowed to hook up to the new lines. Once that fee was paid and a farm received power, there would be a bill sent each month. There would be a charge for the power used, and a small fee for the telephone service.

He continued to explain that the company was ready to start construction and the farms on the map would all have power within six months. All the attendees learned that they would need to sign up during the next week and pay their hook up fee. A representative of the company would be at the town hall every day for the next week.

When he was finished explaining the process, he asked if there were any questions. Most of the hands in the room all shot up at the same time. He called on one farmer who asked how much the hook up fee would be.

Mr. Petters sort of laughed and said, "I guess that is the number one question you each have. I am sorry that I did not mention it earlier." He went on to say that the amount was one hundred dollars. There was total silence as that was a huge amount for many of the farms on those roads.

He said, "Now don't get all worked up. Our company knows that this is a fairly large sum of money for many of you. So, when you sign up this next week our representatives, will work out a payment plan so you do not have to come up with the total fee all at once."

This answer seemed to reassure most of them because he waited for a long time and there were no more questions. He closed the meeting by saying, "This is a great opportunity to bring electricity to the rural people of our county. My company looks forward to working with each of you."

Folks got up and started moving out of the building. There were little groups forming who were all talking about the meeting. We just worked our way out and headed for home without saying anything to anyone.

On the way home Grandpa said, "Well I guess it is time we moved forward and signed up for this program. The prices for corn and pigs are the highest we have seen in many years. The government is buying everything and shipping it to Europe because of this war."

"We can sign up for the payments and then pay it off after Mr. Amos does our harvesting in August. After paying for harvest and the electricity, we will have more money left than we have ever had in previous years. I think we should do it." He finished speaking and Grandma said she totally supported the decision.

It was settled. We were going to have power and a telephone in the next few months. I did not know how much that was going to change our lives.

The fall came quickly. Our corn was harvested and sold. Grandpa paid off the power company. We had sold another lot of our pigs and Grandpa said we had more money than any time he could remember.

He used some of the money to buy lumber from Mr. Scott and barbed wire to repair fences around the barn. The old truck got a new set of tires. Grandma got more canning jars and a new apron. I did not need anything because I was still wearing all the clothes we got from the nice family in town. Life was good.

In December, school was out for the usual Christmas season. Grandpa had picked up a new 1941 Calendar at the feed store. Grandma had already started marking notes on some of the pages. The mornings were cold, and it always took more time for my chores when ice was frozen in the water buckets.

These cold months just sort of slowed things down. There was no baseball practice, so I ended up just throwing a ball or a stick for Ogi. It got dark so early that in the evening we just sat around the wood stove and sometimes grandpa would read to us from the latest newspaper.

The war in Europe was getting worse. Germany had invaded lots of countries. There were a couple of young men in town who had gone over to England to help them fight. It was difficult for me to understand all the things about war. In my little world, people just tried to get along.

I could tell that both Grandpa and Grandma were very worried about the whole war issue. Every evening when he prayed over our dinner, he always asked God to protect us, the country, and the boys fighting in the war.

The days rolled by and before long it was almost Easter. The days were getting warmer. Grandpa was making plans for the spring planting and Grandma was tilling up her garden.

On the way home from school in April, Grandpa said he wanted to show me something. We turned off the road we always took to our farm and went down to the next road to our north.

There were a lot of trucks and men working along the road. Grandpa parked and we walked a few hundred yards up to watch them. On the way he told me they were men from the power company. Before we got much closer, I could see brand new poles that had been placed in the ground every so often.

We stood and watched as they dug a hole with a machine and then lifted the pole up straight and dropped in into the hole. It was very interesting to watch. Grandpa said that after they put up all the poles, other men would come out and place wires on the top of the pole where they would have installed a cross arm.

We watched for quite some time and Grandpa walked over to a workman who seemed to be in charge. He asked the man if he knew when the workers might start installing poles on our road. He was told that it would be about three more weeks.

Grandpa asked him what happens after the pole and wires are installed. The man told him that another crew would come and install a wire from the poles to our house. Grandpa wanted to know how we got lights and places to plug things into.

The man was very patient and explained that there were several companies who followed the construction crews from town to town. He said that people from those companies would contact each farmer and offer to do all the inside installations for a fee. Grandpa thanked him and we walked back to our truck and drove home.

Grandpa told Grandma everything the man from the power company had told us. After dinner Grandpa suggested that we all sit down and discuss a plan for where we would want electricity.

Grandma gave Grandpa a tablet and a pencil so he could take notes. Each of us had a turn of suggesting places to put lights and plugins. Grandpa recorded all our thoughts. In the end we had our plan.

Now we had to wait for the poles and wire to be constructed and then meet with the companies who would do the actual installations in our house. The big question in Grandpa's mind was how much it was going to cost to complete our little plan.

Grandpa assured Grandma that we should be able to pay for all the work because the prices for all farm products were very high as the war continued in Europe.

I believe that we were all excited about getting power to our little farm. It would change so much of our daily life. We could put a pump in our spring and there would be no need to carry water in buckets. I could do chores in the barn even when it was dark outside. We could listen to the radio. Grandma could do sewing and read her books without having to use a lantern.

It was going to be wonderful. But there were dark clouds forming in our future. We just did not know it was happening.

## CHAPTER TEN

The last few days of April, the power company moved men and supplies onto our road and started the same process that we had seen a few weeks before, on our neighbor's road.

On the ride home from school each day, Grandpa and I would each say when we thought they would have all the new poles installed all the way to our house. It became a game between us.

My second year of school was almost over. I was reading much better, and Grandpa had been testing me by having me read parts of the newspaper. He was very patient and would always help me on the words that I did not know.

We still spent a little time almost every evening doing baseball practice. Grandpa said I was getting better every day. He had stopped moving me back from the barn because he was satisfied that I could throw the ball accurately from the distance he had measured out from the tire on the barn.

My hitting skills were also getting better. I was able to hit most of the balls that he pitched to me. Every now and then, he would really throw it hard, and I would swing and miss. He would give a little chuckle and tell me that I still had a lot to learn. But I could tell he was happy with my progress.

Mr. Amos had moved his equipment to our farm and within a few days he had worked up the fields where we were going to plant our corn. He would stop at our house each day on his way home and sit on our porch, where he and Grandpa would discuss the news about the war.

School was just the same. Reading, writing, math and lunch. I still did not play baseball at recess with the other boys. They no longer even asked me to play. Maggie and I ate our lunch and then walked around the playground talking about our families.

Maggie's father owned a lumber mill and cut up logs to supply the town with wood products. Everyone in the entire area liked him and felt that he was a good man.

Grandma knew her mother and they would sometimes meet with other mothers and do quilting. Their family usually bought a couple of pigs from us each year. Sometimes Grandpa would trade Maggie's father pigs for lumber.

I guess it would be fair to say that our families liked each other, and everyone was happy that Maggie and I were friends. She continued her self-appointed role as my mentor in all things about school life. I was more than happy that she was willing to have that place in my life.

We got the rain we had hoped for and a few days later Mr. Amos showed up with his tractor and planter. It only took him two days to plant our corn. Grandpa was very happy.

The progress of bringing electricity to our road, continued at a steady pace. To me it seemed like a lifetime because I was so excited to see it finished. Each day as it got closer to our property, I would walk down the road and watch the last hour of their workday. It made the whole process even more interesting. I got to know the foreman Mr. Anderson, very well. When he would see me coming, he would always say, "How are you today, Mr. Tommy?"

School was out by the end of May. Our corn was growing well and so were the weeds. It seemed for each one that I was able to destroy with my hoe, there were three more that took its place.

Ogi and I always headed for the lake after hoeing. It was

a wonderful place to cool off and play. I had noticed lately that Ogi would tire out sooner than in years before. I do not know how old he was because he had been there my whole life. He was getting a lot of gray hair around his mouth and nose, and it seemed at times that getting up after he had been laying down for a while was becoming a bit difficult.

He still loved to fetch, and it did not matter if it was a stick or my ball. After we played at the lake, his naps in the sun were getting longer. I decided that I would speak with Grandpa about my dog.

I kept up my visit each day with Mr. Anderson. He always seemed glad to see me. The pole line had progressed to within a few hundred yards of our house. As he had told us, another crew was following him down the road. They were putting up a long piece of wood on the top of each pole. These were the cross arms and would become the place where the wires were attached.

Sometimes Grandpa would walk down with me to talk with Mr. Anderson. Those visits became the highlight of my summer days.

By the second week of June, the pole line had reached Mr. Amos' property which was the end of the road. The crossarm crew was right behind them and finished a few days later.

The final phase was to string the wires and attach them to the cross arm. That crew started at the end of June and finished by mid-July.

As Mr. Anderson had told Grandpa, several company representatives stopped by our farm and talked with Grandpa about installation prices. Each time they wanted to see where we wanted power outlets and lights.

Grandpa had become tired of retelling the same story so many times. So, he had created a little map of our property and of the house. It showed every place we wanted power and what the things we would be powering.

After more than three weeks of this process, Grandpa made a deal with one of the companies. When he signed the contract, the company representative told him it would be more than thirty days before they would be able to begin our project.

The summer days slipped by and the corn kept growing. The good news was that the corn was now so tall that the weeds were unable to get sunlight and stopped coming. The day Grandpa told me that the hoeing was done for the year, was the best day ever. Freedom was at hand. I rushed out and told Ogi all about it, but he was less than concerned. He just wanted to play.

One day after my morning chores, Grandpa asked if I would like to go with him down the road to one of our neighbor's homes, where the same company we had contracted with was just finishing up the installation.

We drove down to the Collins house and there were two trucks from the company parked beside their house. Mr. Collins had seen us drive up and came outside to greet us. He and Grandpa shook hands, and we sat down on his porch to talk.

He told Grandpa he was very satisfied with the company and the workers. He asked Grandpa if he would like a tour. We both answered that we would.

We started in the house because it was already completed, and the crew was in his barn finishing up a few last-minute chores. Mr. Collins was obviously very happy with having electricity in his home. He took us through every room, turning on lights and showing us where the plugins were located.

It was amazing to see how the lights worked. At school, the lights were always on, and I had never seen how dark it was without them. This was quite something. Just a little switch and the whole room lit up.

We finished looking through the house and headed out to his barn. The workers were still there but he was able to show us the lights being installed and all the switches. There was a very large light on the front of the barn which Mr. Collins told us was so bright that it was supposed to light up the whole area from the house out to the barn and even as far as the outhouse.

It was hard to believe that could be possible, but he seemed to know all about the lighting. Because these workers were coming to our house next, Grandpa was able to find out from the foreman that they would be starting on our farm the coming Monday morning.

We thanked Mr. Collins for the tour and headed home to tell Grandma the news. We got home and Grandma was out in her garden. She stopped working as we walked toward her, and Grandpa started in telling her everything at once. She said, "Easy there Roy, slow down and tell me one thing at a time." I had never seen Grandpa so excited. It was a side of him that was fun to see.

When he finished and had answered a few questions that she had asked, he said "They will be here Monday Ida!" He grabbed her basket and tools and said, "Let's go to town and get an ice cream." That was the thing Grandpa always did to celebrate. And that is what we did.

By the time we were heading home from town, it was dark as we drove by the Collins house, for the first time in our lives, we saw lights on in their home and on the barn. It was quite a sight. Grandpa said, "In a couple of weeks, our home will look just as nice in the dark."

The next few days were busy. We were moving furniture away from the walls so the men could run their wires and put in plugs. Grandpa had me help him in the barn so there was also room to put in our new lights. We had to move some hay and farm equipment so that a wire could be run out to our spring.

Grandpa had traded Mr. Scott two of our calves born in the spring. We got enough lumber to build a taller platform for our whiskey barrels. Grandpa said this would give us more water pressure inside because we were using gravity to do the work for us.

There were so many new things happening around our farm that it was difficult to imagine just how much all the new changes were going to have on our little family. We were just finishing breakfast when the crew arrived.

They had a copy of Grandpa's little map. The foreman wanted to go over every detail before they started the work. His name was Sam and he and Grandpa toured our property so that Sam could see every detail of the plan we had made.

When they were finished talking, Sam had a meeting with his men. He assigned each one a task and like a well-oiled machine each one headed off with their tools and wire. Grandpa asked Sam if we could watch the workers as they installed our wiring. He said it would be fine.

Each day that they worked on our home was an adventure for me. I had never seen anything like what was happening in our house or our barn. Each of the workers seemed to know exactly what their job was and there was very little talking between them. Once or twice a day Sam had a question for Grandpa. Other than that, the whole crew worked hard, and it was amazing how much they were able to do each day.

After a few days, the newness of what was happening

sort wore off. The crew did the same sequence of things, every day. The only difference was that it was in a different room or area of our house.

On about the fifth day that Sam and his crew were working, Grandpa told me that we were going to start on our new platform for our whiskey barrels. He felt that we would not be in the way, and it was as good a time as any to get it done.

Mr. Scott had a truck bring us out the lumber a few weeks before and it was piled conveniently near our existing platform. That evening Grandpa showed us his plan to remove the old platform and build a brand new one that would be larger and higher. He intended to install several more barrels which would increase our water storage.

Grandma asked him why we needed so much storage, and he just smiled. I think that there was more to his plan than he was showing us. I guess we would just have to wait and see.

We now had wires running from the pole on the road, to our home. And there were wires running from the house out to the barn. The crew had mounted a metal box in our cellar which Grandpa told me was a fuse box. From that box, wires went all over our house and out to the barn.

Most of the crew members were in the house finishing up installing lights and plugins. There were two others putting in a couple of small poles and wires out to our spring. Grandpa said they were going to put another metal box on the last pole and that would be where our new electric pump would get power.

He told me that we would need to bury a pipe from the pump all the way to the new platform we would be building. That was one of the most important changes for me because I would no longer need to be hauling buckets of water from the spring to the house.

It took another week for the men to finish everything on our wish list. Before they left our house to go to Mr. Amos's farm, the foreman walked with Grandpa through every room and turned on every switch to make sure everything was working properly.

When the final inspection was finished, Grandpa asked Sam to give him some pointers on how to hook up the pump at our spring. Sam said, "Roy it would be easier for me to just do it, and I would like to do it for you as a gift for giving us your business." Grandpa was very happy, and it only took an hour or so for Sam to finish that chore.

When he was all done, Sam told Grandpa to flip the switch. When he did, water shot out of the short pipe from the new pump. I do not think that I had ever seen Grandpa so excited. His father had hauled buckets of water to the house until Grandpa was old enough to take his place. That had been his duty until I was old enough to take on that chore.

Our lives were changing very rapidly, and it seemed that it was all for the better. Little did we know that in a few short months, life as we knew it would be forever altered.

Summer was coming to an end. In a short time, I would be back in school. This year I would be in a different classroom with third and fourth grade students.–Grandpa and I had torn down the old platform where our whiskey barrels had been located, for who knows how many years.

We had installed a new pipe from the spring to the house. The new platform was much larger and higher than the old one. The new pipe ran all the way up to the top of one of the barrels. We now had eight barrels, and they were all interconnected with short pieces of pipe. The last pipe ran into the kitchen.

We had lots of water pressure and Grandma was very happy. When Grandpa was satisfied with all our efforts to improve our lives, he told me that we had one last project. After swearing me to keep our secret, he told me we were going to build a little room onto the house. The room would have a shower and a toilet, just like the ones at the school.

When we were finished, we would not need our outhouse, and we would be able to use a shower to bathe rather than having to take a bath.

Our crop looked so good. We had lots of rain during the growing season, and the ears of corn were very large. Grandpa had gone to the bank with a had written crop estimate from Mr. Amos. The banker was able to advance Grandpa money for our secret project. Mr. Scott ordered everything we would need for the job. To help us keep the surprise for Grandma a secret, he only delivered us wood and roofing materials.

With less than three weeks left before school started, Grandpa wanted us to be working on the new room each day after the morning chores. For me it was fun and a great learning experience. For Ogi it was very depressing. There was no time for playing and fetching. No time for baseball practice.

Grandpa was running out of excuses as to why we were adding onto the house. Grandma would no longer accept short evasive answers. Reluctantly Grandpa finally told her it was a present for her and, she would have an indoor toilet and shower. She acted totally surprised, but she would confess in later years that she had pretty much known for the beginning.

A few days before school was to start, Mr. Scott had his truck deliver the shower parts and the new toilet. Grandpa told me that I had been a great helper and that this part of the project would be low and would not be finished before school started.

He told me to take Ogi and go to the lake and enjoy the last couple of days before school would begin. He did not have to tell me twice. Ogi somehow knew that today was going to be different. He would not leave my side, and he kept bringing me sticks. I finally told him we were going to the lake. Grandma made us a large lunch, and we were off to our favorite place.

I forgot to bring my fishing pole, so we swam in the lake and played fetch for a couple of hours. Then we ate our lunch, and both lay down happy and tired.

The sun had moved and what was a nice shady spot, was now half as big as it was when I had gone to sleep. Ogi heard me moving around and we decided to take one last swim before heading home.

The first thing that I did when we got to the house was to check on Grandpa. He had finished the last of the new floor and was just sitting in the shade drinking a glass of tea.

He smiled and asked if we had enjoyed the day off from work. I told him it had been wonderful. Grandpa told me his goal was to at least get the shower roughly installed so I would be able to use it on my first day of school.

At dinner I told Grandpa that I did not need any more days off before school started and that I wanted to help him the next day. He smiled and said, "ok". I could see that he was tired and could really use my help. Grandpa being Grandpa, he did not want to admit that he really could use an extra hand.

I loved this man who had taught me everything that I knew. He was quiet and unselfish, loving and patient. I wanted to be like him when I grew up. In November he would be sixty years old. To me he was young and strong. I finally understood why Grandma called him her oak. The oak trees on our farm were tough and strong. They were able to house birds and critters of every kind. And they produced acorns that grew into oak trees.

## CHAPTER ELEVEN

School started and as planned, I had a shower that first day. It was much better than using a tub. I had grown some over the summer and Grandma had chosen some of the clothes we still had from the Williams family hand-me-downs. They seemed like new clothes to me, and I wanted to look special when I saw Maggie.

This year, we had a new teacher her name was Miss Bandcroft. She was from Texas, and this was her first year, teaching at our school.

After we settled in that day, she told us that we would be learning to write along with improving our printing skills. Math would be learning our times tables and doing division. We would also be learning some history and geography. The newest adventure would be taking a field trip to the library and getting our own library card so that we could check out books to read at home.

It all sounded ok, but I wanted lunch to come so I could see Maggie, and we could talk about the summer adventures. The bell finally rang, and I grabbed my lunch and hurried to walk out to the lunch tables with Maggie.

I was smiling so much my face was hurting. She looked so pretty. Her hair was a little longer and she was taller. I think she was glad to see me as well because she was smiling as much as I was.

We sat at our favorite table and both of us started talking at the same time. We laughed and she was the first to start again. Maggie wanted to know the three best things that happened to me the past summer.

I thought for a moment and said, "Well, we got electricity, installed in our house and barn. We now have a pump that brings water from our spring to our new room where we have a shower and a toilet. But most of all, I got to work with Grandpa on building our new room.

She asked me a few questions about each adventure. After answering all her questions, I took my turn and asked her the three things she like that summer.

Maggie told me that her whole family had gone to visit her father's brother in Texas. While there, they had driven south to see the ocean and the beaches where they stayed for an entire week.

They had built sandcastles and gone swimming in the gulf waters. They had eaten shrimp and all kinds of fresh fish. Her father's brother had chartered a boat, and the captain had taken them on a long ride that lasted an entire day.

She was so excited to tell me all about the trip. I was feeling very plain and sheltered because I had never even been away from our little town. I did not know what a shrimp even looked like let alone have ever eaten one. I had never been on a boat or seen the ocean. Never built a sandcastle and never eaten any fish but the ones from our little lake.

She must have sensed my feelings because she grabbed my hand and said, "I would have much rather gone swimming in your lake. I knew she was trying to make me feel better, but it was not really working.

Maggie was my best friend, but we came from different worlds. Her dad was wealthy compared to Grandpa. She had been places and done things which I had only seen in books at school. I was disappointed because I was unable to relate to her adventures. My life was fine. I never wanted more than what our family had. But this was the first time where I had feelings of being inferior to Maggie.

My life was happy, my grandparents loved me, and I loved them. I had just grown accustomed to having Maggie as my friend. Now there were a few fears that were sneaking into my mind. Maybe she just considered me a charity case and that is why she treated me so kindly.

I made up my mind to stay friends with her but not to make her such an important part of my life. I wanted her to like me for who I was and not what our family had. It was too much for a soon to be nine- year- old to ponder.

When we had almost finished our lunch one of the fourth-grade boys asked me if I wanted to play baseball. I did not really want to, but I was feeling uneasy with Maggie so I told him I would play.

When we got out to the field, it became clear that I was going to be the youngest player. I followed the boy who asked me to play over to a group that was obviously his team. They asked me what position I liked to play. I told them that it did not matter because I had never played in an actual game. It got very quiet and then one of the boys said that I could play center field.

I knew where that position was on the field because Grandpa had taught me a lot about the game. As I was turning to take my place on the field, I noticed Hank Evans was standing on home plate. He mouthed the words "sissy boy" to me and laughed.

I made my way out to center field and then someone said, "play ball" and the game started. Our pitcher through the first pitch to Hank. It was a ball and Hank let it go by the plate. The second pitch was right down the middle and Hank took a big swing. His bat met the ball, and it launched up in the air heading straight for me in center field.

I kept my eye on that ball the way Grandpa had taught me. As it got closer, I could tell it was going to the right. With all the speed that I could muster, I ran to the right and was able to reach out and catch the ball. My hand sort of hurt because I had never owned a baseball glove. Grandpa and I just used our hands when we were playing. It was so fun to catch a ball bare handed that had been hit by the school bully.

The pain was overshadowed by my joy of catching that ball. I looked over at Hank. He was angry, very angry. He walked back to his team and threw his bat against the fence.

We got two more outs and then it was our time to be at bat. Several of the players on my team had expressed their delight that I had caught the ball hit by Hank.

The boy that seemed in charge of our team was a fifth-grade boy, named Todd. He came over to me and said, "That was a great catch, where is your glove?" I told him I had never owned a glove. Then he asked me if I could hit as well as catch the ball. I told him that I was ok at hitting. He said, "Alright let's see what you can do. You will hit second."

While Todd was talking to me, I failed to notice that Hank Evans had taken the mound, and he was the pitcher. Our first batter had hit a pop-up fly ball that the shortstop had caught for the first out.

When I came up to the plate, Hank looked meaner than usual. He wound up for the first pitch and threw it right at me. If I had been any slower, it would have hit me right in the head. Our team started yelling at him to knock off the dirty throws. He just glared back at them and did the same thing on the second pitch.

We had a couple of older boys who were about Hank's size, and they ran out to the mound and there some pushing and shoving going on for a minute or two.

When they walked back to our side. Hank looked a little less inclined to knock my head off with the ball. He through a low pitch for ball three. Then he threw two strikes that went right over the plate. I was used to how Grandpa tossed the ball, and this was new to me.

On the next pitch it came sailing right over the plate as the last two had. I was ready this time. My eye was on that ball, and I made a hard swing. There was a noise when the bat met that ball and it flew high and long toward left field. Because I was smaller than most of the players the outfielders were playing close to the infield. The ball flew over the left fielder and landed on the grass and proceeded to roll and bounce toward the back fence.

I ran as hard as I could, rounded first base and headed for second. By the time I hit second base the fielder had still not been able to grab the ball for what was a long throw to third base. I kept running toward third. The shortstop had picked up the ball thrown from the outfield and threw it toward third. The third baseman missed the catch, and I ran for home plate. The ball never got there in time. I had scored a homerun.

The rest of the game was not as exciting to me as my first experience of the ultimate "homerun." We ended up losing the game because Hank's team was ahead when the bell rang, signaling the end of recess.

I did not hear much of what Miss Bandcroft had to say that afternoon. I was reliving the first game of baseball I had ever played. Grandpa was going to hear the whole story on our way home.

The bell finally rang. Maggie came up to me and asked if everything was ok. I told her everything was good. She looked a little disappointed that I did not have more to say.

I told her that Grandpa was waiting, and I would see her the next day. On the way home I was so excited that all my words ran together. Grandpa just smiled as I related every single detail about the ball game. The only thing that changed that smile was when I explained how Hank had tried to hit me on the first two pitches.

When we got out of our truck at home, Grandpa came over to me and put his arms around me and said, "Gule (acorn) I am very proud of you. You have learned your baseball skills well. You have also made an enemy. Hank will never be your friend, because you have shamed him in front of the other boys. He is tla osda (no good). You must be on guard around him. He will try and get even with you." I did not know that I was soon to learn what Grandpa had foretold.

I did not want to go to school the next day. The thought of seeing Maggie again, was making me uncomfortable. I did not know what I would say to her. It was though she was now a stranger to me.

I know it was crazy to think that way, but it was difficult for me. I finally fell asleep and woke up still tired when Grandma told me breakfast was ready.

When we arrived at school, Grandpa said, "gule (acorn), I love you and it pleases me more than you can know, to watch you grow and turn into an asgaya (man)." He gave me a hug and drove off.

As I was walking toward the school door, Hank and a couple of his buddies approached me. I just kept walking. Hank blocked my way into the school and told me that I would be playing baseball again today.

He said, "No sissy boy is going to get lucky like you did yesterday." Then he shoved me through the door, almost knocking me over.

Getting no response from me seemed to make him even more angry, so he pushed me again. The second push made me fall. My lunch bag ripped open and spread out on the floor. He took his foot and squashed every item that had been in my bag. Then he and his two friends laughed at me.

As I was trying to get up, I saw Miss Brandcroft coming down the hallway. Hank must have seen her to because they quit laughing and moved on down toward their classroom. Miss Brandcroft asked me what had happened. I did not know what to say, so I told her that I had tripped.

She looked at my lunch all squashed into mush and said, "Tommy, do you have something you would like to tell me?" I just shook my head and kept looking at the ground. She helped me clean up my lunch and we put it in the trash. As we walked to our classroom, she told me, "This is not over Tommy, but for now it will stay between you and me." I just nodded my head.

The morning was like torture. I knew I would have to see Maggie at lunch and try and explain why that I did not have any lunch. I knew Hank was going to make a big thing about me playing baseball at recess. I was miserable.

When the lunch bell rang, I could see Maggie heading toward me with her lunch. I tried to smile at her. Her first comment was, "Where is your lunch?" I said that I was not very hungry. That comment was not going to satisfy her. She said, "But where is your lunch?" This was not going to work out well, so I told her that I had dropped it coming into the building. I was unable to look at her, so my head was just hanging down.

"Tommy, that does not make any sense," she said. I asked her if we could just sit and not talk about it. It was very difficult for her, but she honored my request. It was a quiet lunch with her offering to share her food with me.

Just before the bell rang for recess, Todd came to our table. He said, "Tommy, this is a baseball glove that I had a few years ago, I never use it, and I want you to have it. We really liked having you on our team yesterday and hope you will play with us again today."

That was the first happy thing that had happened to me the entire day. I assured him that I would play with them today and I said wado (thank you) for the glove.

The recess bell rang, and all the players moved out to the baseball field. Hank announced that the teams would have the same players as the day before. Todd told him that he was not in charge but that we would go along with the plan.

It was decided that Hank and his team would bat first. Todd was assigning players to positions and when he came to me, he said, "Well you can hit, you can throw, can you pitch?" I told him that I could. So, he agreed to let me start as the pitcher.

The first batter was one of the boys who was with Hank that morning. He stood at home plate and sneered at me. Then he made a motion with his arms as though he was rocking a baby. I had seen him before in the sixth-grade classroom.

I stood on the mound and pretended I was looking at the tire Grandpa had nailed to the barn. That same tire that I could through a hundred balls into without missing.

I threw three perfect strikes that he tried to hit each time. It was the first out. The next batter was a boy that I had never seen. I threw two perfect strikes. He never even tried to hit the ball. The third pitch was perfect. He hit it but it went straight to the first baseman, and he was tagged out. Out number two. The next batter was Hank Evans.

He walked up to the plate and pretended to throw me a kiss. I decided to throw the ball as hard as I could. The first pitch was a swing and a miss. Hank looked a little less confident. The second ball was a swing and a miss. I waited for him to get ready, and I threw the fastest ball I had ever thrown. By the time he made a swing at the ball, the catcher had it in his glove. Out number three. Hank walked back to his team. His expression was one of defeat. He did not look so tough now. His world had been turned upside down by a third grader.

My team was yelling and cheering. I had slaps on my back, hugs, and pats on my head. Todd came up to me and said, "Tommy where did you learn to pitch like that?" I told him about Grandpa and the tire nailed to the barn. He just shook his head and smiled at me.

The rest of the game went pretty much the same way. The batters were unable to hit the balls that I was throwing. Our team was able to score two runs before the bell rang. Each team had won one game this week. I guess there would be at least one more game to determine who had the best team.

As I walked back to our classroom, I could see Maggie watching me. I felt bad that we were not as close as we were a week before. It was something that I did not understand, and it made me sad.

My little life was becoming more complicated. Electricity, bullies, water pump, indoor toilet, shower, and relationships were all new to me. My entire life had been Grandpa, Grandma, and Ogi. There were a few moments when I wanted it to go back to what had always made me happy.

When Grandpa picked me up, I related everything about the entire baseball game to him. I left out any mention of losing my lunch. I showed him the glove Todd had given me. Grandpa turned it over in his hands and commented that it was a quality glove and would last me for a very long time.

There was no talk about Maggie. That was not something that I wanted to share. Grandpa could see that winning the game made me happy, but he was also able to see an underlying sadness. He gave me space and did not force the subject. Grandpa was a very wise man. He carefully thought before he spoke and when he did speak, it was focused and direct.

When we got home and were getting out of the truck, he reached his arm around me and said, "Gule" when you are ready, you can tell me what is weighing heavy on your heart."

What a gentle and understanding grandpa. I was very blessed. He was not just my grandpa he was my best friend. He was my tala (oak) and I was his gule.

At dinner Grandma wanted to hear all about the ball game. She was very excited that all the lessons Grandpa had given me were helping me to be a good player.

Grandma announced that on Saturday we were all going to town to visit an appliance store. That was a word I did not recognize, so she explained that because we now had electricity, we could purchase some items which were called appliances. They were things like a toaster for the bread, a clock that we did not need to wind, and a waffle iron.

That sounded like a fun adventure. There were still a few days of school before Saturday. I was not looking forward to being confronted by Hank Evans. Maggie was also in my thoughts. Sleep did not come easy that night. There were too many things on my mind.

When I finished my morning chores Grandma had breakfast ready and my lunched packed. We were running a little late, so I ate my bacon and biscuit in the truck.

I never knew why, but Grandpa parked the truck and walked me up to the school door. Hank and his buddies were waiting. When they saw Grandpa, they moved aside.

Grandpa roughed up my hair with his huge hand and said, "see you after school gule." I walked to my class and put my lunch and my glove on the shelf next to where Maggie kept her lunch.

Miss Bandcroft was already sitting at her desk. She watched me walk in and said, "Good morning, Tommy". I was about to take my seat when she asked me to come up to her desk. I walked toward her with my head down because I had never been asked to come to her desk. Maybe I was in trouble. There were lots of things running through my head. She smiled at me and asked, "Did you have any problems with your lunch this morning?" I said, "No I did not". Her only reply was, "Good".

All morning, I kept wondering, why had Grandpa walked me to the door? Why did the teacher ask me if I had any trouble with my lunch? Did she say something to Grandpa about what Hank had done to my lunch?

We spent the morning working on math and writing. The lunch bell rang, and I grabbed my lunch at the exact time that Maggie picked up hers. She smiled and, ask me if we were eating together. I told her that it would be fine.

We sat at our usual spot and neither of us said anything for a long time. Finally, Maggie said, "Tommy, why are you avoiding talking to me?" That was a direct question, that I was not prepared to answer.

I pretended to chew my food so I could stall and come up with something to say. After a minute or so, I said "Maggie, I am just a farm boy and there are so many things that you know and have experienced. I just feel like we live in different worlds."

"My world is a small farm with my grandma, grandpa and my dog, Ogi. I can tell you about corn and cows, about chickens and hogs. But I do not know much beyond those things. You live in the town. Your home has all the comfort things. You have never used an outhouse. You never had to read from a kerosene lamp."

"My shoes and my clothes are hand-me-downs from some nice folks in town. I have never owned a new piece of clothing."

"My skills are milking cows, feeding pigs, gathering eggs and playing baseball. I never realized how different we are until you told me about your summer." Maggie looked very sad, and I think that there was a tear in one of her eyes. "Oh Tommy, I am so sorry if I made you feel bad. You are the most tender- hearted boy that I know. You do not need to feel bad about being a farm boy."

After she spoke it was totally quiet. She grabbed my hand and held it to her face. It was obvious she was crying. When she was able to speak, she said, " Tommy please do not think poorly of me. That would break my heart. You are my best friend. I like who you are. I do not care about your clothes or what you own. I care about you."

I am not sure that it changed all my feelings, but it did make me feel as though our friendship was better for having our little talk.

Our moment in time was interrupted by Todd when he came over and asked me, if I was playing in the game. I told him yes and that I needed to run and get my glove. When I got to my little cubby on the shelf, my glove was gone. After looking everywhere, it was apparent that someone had taken it.

I walked out to where Todd was standing, and he could see that something was wrong. I just shrugged my shoulders and told him that someone took the glove. He was obviously angry. Afterall, it had been his glove before he gave it to me. Todd looked around the field and saw Hank and his two buddies watching us and laughing.

He did not hesitate. Walking very quickly, he got to where they were standing. I could not hear everything that was said. I did hear, "it better be back on his shelf by morning".

Todd walked back to me and asked if I still wanted to play without a glove. I thought about it for a minute and said, "you better believe it." He nodded his head, patted me on my back and said, "Here is the ball and you are the pitcher. Make them pay Tommy."

The game started with our team batting. We had a couple of hits, but no runs. Hank's team had different batters leading off from the ones the day before. It was no problem for me. I struck out the first two batters. Hank was the next one up. He smiled at me and yelled, "Where is your glove farm boy."

That made something in me turn into rage. I threw that ball so hard that Hank was still standing there after the catcher threw the ball back to me. He looked stunned. I did the same thing two more times and he was out on three strikes. My team cheered as he walked away dragging his bat.

The rest of the game was much the same. Every so often a batter would contact the ball, but they were mostly hit into foul territory. Our team finally had a few hits, and we managed to get one run.

By the time Hank was at bat again, his team was becoming very frustrated. They were just unable to hit the balls that I was throwing. This time it was apparent that he was giving everything he had to try and hit one of my pitches. Before I could throw the ball, he yelled at me, "Give it your best you dumb little gule."

I obliged him and threw a strike that he missed. The second pitch he managed to get a little piece of the ball and hit it foul. The third pitch was a swing and a miss. He was getting angrier with every pitch. The next ball was a bit low and a bit slower. He was able to hit that one. I just sort of rolled up to the pitcher mound. I reached down, grabbed it and threw it to first base and he was out.

The bell sounded and we won the game. Todd came over to me and the first thing he said was, "What does gule mean?" I told him that my grandpa called me gule, and that it meant acorn in the Cherokee language. He smiled and told me I had done very well today.

The news of the baseball game was getting around the school and when I got settled in class, I could see some of the kids whispering and looking at me. Maggie smiled at me and gave me a thumbs up.

The afternoon was pretty much business as usual. It went by quickly and as we walked out Maggie came over and grabbed my hand and said, "I am proud of you gule." From that day on, my friends called me gule and strangers called me Tommy.

On the way home, I told Grandpa about the missing glove and people calling me gule. He said we should wait and see if the glove showed up in the morning. He told me I should be proud that they like my Cherokee name.

The next morning when I arrived at school, it was not impossible to wait, and I ran into the classroom to see if the glove was returned. It was not there. Before the bell rang for school to start, Todd walked in and saw me standing by the self. As he approached, I shook my head no. He did not say anything. He patted me on the back and walked out.

By now, everyone in the school knew about the glove and the deadline Todd had given for the glove to show up. By lunch time the entire school knew the deadline had come and gone and no glove appeared.

I could play baseball without a glove. But that was not the issue. Someone had stolen it, and I wanted it back. The prime suspect was Hank Evans or one of his buddies, who did whatever he told them to do.

All during the morning, I thought of ways to get even with Hank or whoever took my glove. Reality finally overcame my anger, and I realized that I was a third grader dealing with boys who were much older and much bigger than me.

There was not much that could be done. There was no proof of who took it, even though most of us had a good idea of who was guilty.

Lunch time came and Maggie and I just sat and talked. It was very nice to visit with her. When the recess bell rang Todd showed up to see if I wanted to play. I told him that I was not really feeling like playing that day. He seemed to understand and gave me one of his pats on my back and walked off.

I did not even watch the game. Maggie and I talked about the glove and lots of other things. I enjoyed the time with her. Just before the bell rang, she asked me if I had another glove, would I like to play baseball again. I told her I probably would, but it was not that important to me.

When school ended that day, Miss Bandcroft asked me to stay with her for a little while. When all the other students were gone, she came over and sat at the desk next to me.

She was smiling at me as she said, "Tommy, you are a very interesting little boy. I have watched you this year and I am pleased to say you are one of the nicest children in my class. I wanted to ask you about the missing glove." I told her all about it and even told her that it was probably Hank Evans who took it.

She waited for a moment and said, "I believe you are most likely correct. I too believe it was Hank. However, we cannot prove it so unfortunately, we just need to wait and see. I do not want you to do anything that will bring you any trouble. Hank is a lot older and not a very nice person." I assured her that I was going to be myself and ignore him and his friends. She agreed that it would be the best course of action.

Grandpa was waiting in the truck, and I used the ride home to tell him all that had happened that day. He did not say much which was not at all surprising. That evening before I went to bed, he said, Gule, there are men and there are boys. You are behaving like the asgaya (man). Do not let the evil done by others, ever change you.

I went to sleep pondering his words. Grandpa was the wisest man that I knew. He was my hero. He would remain so the rest of my life.

I did not play baseball the rest of the week even though Todd asked me to each day. It just was not important to me. Maggie and I were enjoying just talking and laughing. We were becoming better friends each day. I learned more about her, and I was even sharing more about myself. They were wonderful days.

Hank and his friends still walked by our table and said mean things. Their favorite was that I was a spoiled little crybaby and would not play because someone took my glove. I just pretended that they were not even alive. That seemed to spoil their fun and they would eventually walk away.

Each day when recess was over, Todd would say hello and tell me who won the game. He really was a very nice person. He assured me that when I was ready the team wanted me back. I always thanked him.

On Saturday, we went to the appliance store, and we purchased, a clock, a waffle iron, a radio and the biggest surprise of all was a brand-new refrigerator. Grandma was so happy. She just could not stop smiling. "Roy, can we afford this?" He assured her that it would all work out and that she deserved it.

The following Monday, Maggie walked into class carrying a brown box. She put it on her shelf and came over to say hi to me. It seemed that she was extra happy today. I was happy to see her, so we were both smiling. Class started and the morning worked its way until the lunch bell began to ring. Maggie asked me to help carry her lunch and the box out to our table.

As we were eating, I asked her what was in the box. She smiled and said I would have to wait until we had finished our lunch. My curiosity was now growing. When we had both finished, she handed me the box. "This is for you from my father."

I peeled off the brown paper and opened the box. There inside was a new baseball glove. It smelled of new leather. She said, "Try it on." I did and it was just a little big. That was a good thing because I would grow into it. As I was turning it over and admiring everything about my new glove, my eyes fell on a name burned into the leather. It said Gule in big letters.

I did not know what to say. I just stared at my name. Maggie finally broke the spell and said, "My father drove out to see your grandpa while we were at school last week. He told my dad the kind of glove you would be able to use. My dad bought it for you."

I could not find any words. No one outside of my family had ever done anything like that for me. After the knot in my throat went down, I managed to say, "Wado (thank you) Maggie." She asked me what that meant and I told her it was a word that expressed a heartfelt thank you.

I was still overcome by the generosity of her family when Todd came up and asked his usual question of, did I want to play today. I showed him my new glove and said, “yep”.

Hank was pitching and our first batter hit a single. Our second batter walked, and the third batter struck out. Todd hit next and got a single. The bases were now loaded, and Todd made a signal for me to bat next.

Hank looked at me and I knew that he was going to throw the ball and try and hit me just like the first game we had played. I was ready and was able to duck out of the way. Our team started screaming at him and he just stood there and laughed.

For some reason it did not rattle me and when his second pitch was a little slower, I slammed it over the third baseman’s head. It rolled down the baseline all the way to the fence.

The fielder finally grabbed it and threw it to second base. It was too late. We had scored three runs, and I was on third base. The next batter was hit by accident and walked.

One of our best players named Mike was up next and he hit the first pitch Hank threw and it went right past the shortstop. I scored and Mike made it to second. Hank’s team was getting upset with his pitching and the more they prodded him, the worse his pitching was getting. We ended up winning the game six to nothing. Hank was so mad the next day that he and his team refused to play.

Hank was very jealous of me, and it was displayed by his actual hatred of me. Maggie had watched the entire game and was waiting for me when we headed back into the school. She smiled at me and said, “I guess your new glove has caused a stir.”

At the end of the day, Miss Bandcroft came up to me and said, “Well Tommy, sometimes these things just work out. Congratulations on your new glove.” In all the years that I played with that glove, I never left it anywhere. It was always with me.

On the way home I told Grandpa about the game and showed him the new glove that he had suggested to Mr. Scott. He looked at the word Gule, burned into the glove and smiled. "It looks really good." He was happy and a little sad at the same time. I think that he wished that he had been able to buy that glove for me. I hugged him and said, " wado, Grandpa."

# CHAPTER TWELVE

In late September, our corn was harvested. Grandpa paid Mr. Amos his share and then paid our bill at Mr. Scott's. We still owed a little at the appliance store and Grandpa settled that account as well. We still had some money left to help us through the winter.

Life was good. Grandma had her new refrigerator and that made her life so much easier. Grandpa listened to his radio every evening. There was more news about the war, and it became the main topic between Grandpa and his friends.

By Thanksgiving, we had snow and there was no way to play baseball. We built snowmen and threw snowballs during recess.

Ogi always waited for the truck to bring me home from school. He would be waiting with a stick or a baseball at his feet. When he saw the truck coming, he would grab whatever he had and run to the truck. He loved the snow and would play fetch with me for as long as I would throw something to him.

The evenings were a family time. Grandpa's radio would play. I would usually be reading a book. Grandma would be sewing on her latest project. Our fire was warm and added a nice smell to our home. It was a peaceful and safe place.

The school had reopened on December 1st, and we were happy to be back from our week-long break. Maggie and I talked about everything we had done while school was out. I told her that I wanted to do something for her father to show him how much I appreciated the glove he had given me.

She suggested that I write him a letter and that she would deliver it to him for me. I thought it was a great idea, and she even told me that she would help me write it. Maggie was a wonderful friend.

The next day at lunch we talked about what the letter should say and during recess we wrote the letter. When we were finished, Maggie read it aloud and we both thought that it was a good letter. We folded it up and she put it in her lunch box and said she would give it to him that evening.

At lunch the next day, Maggie told me that her father really like the letter and had told her to tell me thank you. I felt so happy. I wished that I had written him earlier, it just seemed that it was not the right time until now.

On Saturday, Grandpa and I worked on some fence repairs and started work on an old tank that he had been given. He wanted to make it into a storage tank for gasoline so that he did not always have to buy gas when he went to town.

We finished our evening chores and had the usual evening of radio, books, and sewing. There was a large pile of firewood just outside our front door and we always brought a few pieces inside so that if anyone got up in the night, they could throw a piece or two into our stove.

On Sunday we had a big snowfall and Grandpa thought we should stay home from church. We finished the morning chores and went inside for breakfast. Grandpa turned on his radio and the news was terrible. That morning, the Japanese navy and their planes had bombed Pearl Harbor in Hawaii. We had lost ships, airplanes and men. The President was even on the radio saying that we were now at war with Japan.

Grandpa did not move at all. Grandma was dabbing her eyes with her apron. It would take a few days to realize the total impact of what we heard on the radio.

After the initial news was over, Grandpa said, "We better

make a list of things we need to do because of this war." Grandpa got a tablet and a pencil, and we started listing things that we would need and things that we already had. We would need gasoline for our new tank, extra sugar, flour and cornmeal.

Even though we now had electricity Grandpa thought we needed some more kerosine, in case the electricity got shut off. We had lots of extra eggs, milk, cream, pigs, a few calves. We could use these to trade for things we needed, or we could sell those products for extra money.

Grandpa decided that we should stock up on pig feed, and chicken scratch. He said he would order a delivery of gasoline in the morning. I did not know that being at war would impact us in so many ways. Grandpa and Grandma had lived through World War One and they seemed to know what might be happening in our future. Grandpa surprised us both when he said that he was going to order us a phone. He felt we needed to be able to talk to our neighbors in case there was an emergency.

Grandpa decided that I would not go to school the next morning. He said he would go to town, and I could go with him and help with supplies. When I headed off to bed, they were still talking about the war.

Monday morning, when I came in from my chores, Grandma had breakfast ready, and Grandpa was listening to the radio. I do not think that either of them slept very well because they both looked tired and worried.

We ate without any conversation and when we were finished, Grandpa got the list he made the night before and we got in the truck and headed for town.

Our first stop was the town bulk fuel supplier. Grandpa told him we had a new one-hundred-gallon tank and paid him for the delivery. The owner told Grandpa that it was a good thing to order everything we needed now because fuel would no doubt become hard to get as time went along. Grandpa asked him to deliver four, five- gallon containers of kerosine as well.

The next stop was the feed store. Grandpa ordered 30 sacks of pig feed. Each one was fifty pounds. We had to take them home because our truck did not carry much more weight. After placing them in the barn, we headed back to the feed store and got thirty sacks of chicken scratch. Those were also fifty pounds each. We had to make another trip home to the barn.

We had lunch before heading back to town for more supplies. This trip was to the telephone company where Grandpa signed some papers and ordered our new phone. He was told that it would be about three weeks before it would be installed.

The next stop was the mill. They sold flour, corn meal, and sugar. We filled the truck to almost full, and Grandpa remembered that we did not get any salt licks for the cows. By the time we added those our truck was loaded, and we headed for home.

By the time that Grandpa and I had put everything away and done the evening chores, we were both tired. That was ok with me because I had a wonderful time with Grandpa that day.

After dinner, Grandma and Grandpa listened to the news on his radio. The news was no better. Many people had died in the bombing, and several ships were destroyed or badly damaged. The government was asking for volunteers to sign up for the Army, Navy and the Marines.

Grandma had spent the day making a list for Grandpa. It seemed that many of the folks were stocking up as we were trying to do. We bought candles, salt and pepper along with boxes of matches and soap for washing clothes, dishes and for us when we bathed.

Grandma also wanted extra baking soda and baking powder and coffee. As we were leaving for town, she yelled out the door for us not to forget to get lots of vinegar so she could make her famous pickles.

The store still had most of the usual items, so we were able to get everything Grandma wanted along with some extra items that Grandpa thought she might need.

I helped Grandpa put away everything we had purchased in town. He told me, " Wado for all the help." Then he said, "I think you need to go play with Ogi. He is probably feeling neglected. Be home for your evening chores."

That was music to my ears. I went outside and called him, and he came running with a stick in his mouth. I took the stick and told him that even though it was cold, we were going to the lake. He seemed to understand and off he ran in the correct direction. We spent the afternoon playing fetch and walking. It was too cold for taking a nap and neither of us was that tired anyway.

I returned to school with only two days left in the week. The general mood of everyone was very different. The teachers did not seem to know what to say. The children did not understand what war meant. Their parents had not known how to explain everything that was happening. We all somehow got through those two days and when home for the weekend.

Every evening was the same, when dinner was over, Grandpa turned on the radio, and we listened to the latest news about the war. Grandma very seldom made a comment. She worked on whatever project she could occupy her time, while listening to the news. Grandpa would read the newspaper, and I usually read a book.

Grandpa would sometimes tell Grandma about some things he had heard in town. Mostly, we just listened to the radio. It was too cold to do much outside, and our wood stove made for a cozy place to spend the evening.

My life was happy. I loved Grandpa and Grandma with my whole heart. Our little home was warm and cozy. It was so often filled with wonderful smells from the meals Grandma was making. I no longer had to make the trip to the outhouse. We had a shower and electricity. When I was snuggled up in bed and could see my baseball glove on my dresser. Each time that I put it there, I would place it so that the word Gule was facing my bed. In my little world osda nigada (all is good).

As the days of December crawled toward Christmas, almost every part of our lives was about the war. The news, conversations between friends, school lessons, and the posters that started showing up all over town, were just a few of the constant reminders that life as it had been for the last two decades was all changing.

Almost every time that Grandpa returned from town, he would tell us that another of the men who we knew had joined up to go fight for America. There were even some of the high school boys who had their parents sign a paper which allowed them to join without being eighteen years old.

Even though our town was small, a recruiter from the Army had come to sign up volunteers. Life was becoming different far too soon. Little did we know how much it would change.

By Christmas, there were a few items that were no longer readily available at our store. We had people from town coming out to our farm to buy, eggs, milk, cream, and bacon.

We always had a little smoke house to cure bacon and hams from our hogs. The pressure from folks in town wanting more meat, had pushed us to enlarge our smoke house. In the years that followed, I could never remember a time when that smoke house was ever empty. As soon a one batch was finished, we had another one ready to take its place.

We were truly blessed. Grandma started selling chicken pies and, later added pork pies as well. Our chicken coop had been enlarged three times by Christmas of 1942.

For the last year, our hogs supplied food for many people in town. Mr. Amos was now planting and harvesting our corn crop each year. That left Grandpa and me the time to take care of the cows, chickens, and pigs. Our lives were very busy.

School during those years was different. So many of the high school students were busy with jobs that required them to miss school. Several of them had joined the army, including the oldest Evans boy who had been in jail. Apparently, the government would allow some prisoners to leave jail if they agreed to go to war.

I often missed school to help Grandpa with all that he was doing. He had even taught me how to drive our old truck because there were times that we needed to make a delivery to town, and he was not available. The sheriff turned his head from things like this because the times called for unusual activity.

One of the high school students was driving the school bus. Many of the students had fathers who were gone fighting in faraway places around the world. Most mothers had to take on the role of mother and father. Everyone was forced to grow up sooner and to help carry the family load.

By the summer of 1942, the post office had become a

place where people met each other. They were there to mail a letter to a loved one fighting somewhere or there to hopefully get a letter from their family members.

Some left disappointed and a lucky few received a letter. These were also the days when an army vehicle would show up in town bringing the news that some family had lost a son or father in the war.

Magie was busy working with her father. She often missed school just like the rest of us. On those days where we both attended, we tried to pretend that things were normal. Neither of us was living a normal life. All our lives were on hold. Lives would be changed forever by this war.

We tried to keep each other aware of what we were doing. Our friendship continued but it had many interruptions through those war years.

During 1942 there were ration cards issued. You were only allowed to buy a certain amount of gasoline, tires, butter, flour, canned milk, and many other products. Beef and pork were in short supply and that was one of the reasons that our animals were so valuable. All those products were going to feed the troops and other people affected by the war.

My education was not coming from school. It was from working with Grandpa. Math from making orders for people. Logistics from getting deliveries to the correct destination. Learning to slaughter animals and butchering them to supply the various needs of our customers. I learned the best ways to smoke meat to produce a wonderful tasting end-product. We kept records to know how many eggs our hens would produce each day. We used the hens that were older for meat. It was more practical math than I could have ever learned in school.

Our cows were milked twice each day and our customers

needed to pick up their order promptly because we did not have a means of refrigerating large amounts of milk and cream. Each day dozens of folks came from town to pick up milk, eggs and meat.

Grandpa and I were so busy that I rarely attended school. It was sometimes weeks at a time that I never saw Maggie. When I did, it was usually because her family came to pick up a meat order.

Our little barn now had four refrigerators and one large freezer that were all purchased through the appliance store in town. They were necessary to keep the meat fresh until the customers could pick it up. Grandpa finally made up a calendar of when different animals would be ready, and folks needed to make a commitment as to the amount of of meat they needed.

It was amazing how well his system worked. Eventually we had a list for eggs, bacon and Grandma's meat pies. Beef was the commodity that we could not keep up with the demand. Some folks were eating horse meat as a substitute for the beef.

Bacon, ham and sausage were the main stay meats. Grandpa had purchased a meat grinder from Tulsa, and we were grinding our own blend of pork and seasoning to produce sausage. It was one of our best-selling products.

Our little farm was helping lots of people endure the hardships of this war. As time went by fewer families were unable to purchase the staples of everyday life, so they came to us. Grandpa and I were working eighteen-hour days. I almost never went to school.

Some of the other farmers who owned cattle and hogs, milk, eggs and meat were coming to Grandpa to sell him their animals.

Most of them were not interested in slaughtering and

selling meat. It did not take long before we could no longer expand. We just did not have enough refrigeration.

One afternoon we got an unexpected visit from Maggie's father. Grandpa showed him around our little operation while I just kept working. Andy Scott had seen parts of the process before, when he came to pick up his order. But this time he got the full tour.

After an hour of seeing every part of our little business, he told Grandpa that he wanted to talk business with him. They headed off to the house and left me to my duties.

I would later learn that Andy had told Grandpa that this war looked like it was going to last longer than anyone had figured. He felt strongly that the need to supply meat was going to be very important for quite some time.

He told Grandpa that he had watched our little farm operation turn into a business that was filling a local need. Andy told Grandpa that he wanted to go into business with him and take what he was doing to the next level.

Mr. Scott must have spent a lot of time thinking of a plan because Grandpa told us that the plan Andy had come up with was huge and complicated but doable. Grandpa explained to Grandma and me that the basic idea was that we owned the land and already had a business that was working.

Andy had the building materials and the manpower to build something on our land that would make the processing business much larger. On a good week Grandpa and I could butcher and process one beef and two hogs.

Mr. Scott wanted to add much larger refrigeration rooms and a huge freezing room. His idea was to fully process ten cattle every day and twenty hogs.

While Grandpa was telling us the plan, Grandma and I were completely and totally overwhelmed. We could not envision that large of an enterprise. As Grandpa continued with the story, it slowly made more sense. We would employ workers from town. Much of our production would go to feed people in nearby communities. The local cattleman and hog growers in the surrounding area would be able to sell their animals locally and get very good prices for them. Andy thought it was a great idea. Grandma wanted to know what Grandpa thought of the idea.

He took his usual time before replying to her request. "To be honest it will be a lot of work. Tommy and I are already working very hard. I can see how it would meet a lot of needs in town and the surrounding communities."

Grandma pushed harder and wanted to know if it would work without destroying our lives. After some thought he said " Andy is younger, and he has ideas that are much bigger than most folks. He is willing to help make this into a large operation. The truth is, he doesn't even need us. He could do this whole thing on his own."

"He is an honest man, and he will help Tommy and me build this operation. He also wants to employ more women from town to help make and sell your pies all around the area."

This was something that we had not heard about until this very minute. Grandma was looking like she was going to faint. Grandma wanted to know when all this was supposed to happen. Grandpa said Mr. Scott could have the buildings constructed and completely ready for all the machinery and freezers, in less than six months.

Grandma asked him what he had told Mr. Scott. Grandpa said, "Well he and his family are coming for dinner on Sunday, and we are all going to talk together as families".

I was excited. I was going to get to spend some time with Maggie. This whole plan was a bit unreal to me. There was no doubt in my mind, that Mr. Scott and Grandpa could make something work.

We were so busy filling orders, cutting meat, feeding our animals, and making deliveries, that there was no time to do anything but work and sleep. Grandpa did not show any indication of slowing down. It seemed that the success of our little business was a fountain of youth for him.

## CHAPTER THIRTEEN

On Sunday morning, we went to church and when it was over, we hurried home to get ready for the Scott's visit. Grandma had made one of her now famous pork pies, some deviled eggs, fresh cucumber salad from our garden and buttermilk biscuits.

Grandpa and I helped her set the table so that everything would be ready for our guests. A few minutes after one o'clock, they arrived. Grandpa welcomed them into the house. Andy carried in a large tablet that he laid on the sofa as we sat down at the table.

There was lots of small talk and then compliments about Grandma's cooking. When everyone was full, we cleared off the table and Andy retrieved the tablet from the sofa and brought it to the table.

Mr. Scott opened it to the first page and placed it in the middle of the table. Then he began to explain the drawings and comments on each page as he worked his way through the tablet.

He complimented Grandpa on how he had been able to build up a business that was helping feed our community. He went on to explain how the war was going to put more pressure on the food supply and limit local people in their ability to have fresh meat, eggs, milk and many other products. He explained that to expand processing and distribution, that we would need a lot more electricity.

Anticipating this need, he had contacted some government people about our business, and they had told him that if we were to supply the armed forces with sixty percent of our production, that the government would have the needed electricity sent to our farm and not only that but they would assure us of the necessary licenses and permits we would need to open a new plant.

Andy told us that the government agent also told him that any equipment that we needed would be located and delivered by the government and that fifty percent of the cost would be paid for by the United States military.

Grandpa was not prepared for all this information. He looked a little dazed as Andy continued to speak. The agent had told him that the government needed this type of business to help provide supplies for the troops. When Grandpa could finally speak, he said "Andy, is this too big for our little town?"

Mr. Scott looked straight at Grandpa and said, "Roy, you started something that will be of great value to the government and to the people of our community. With the government's help and money, you and I are going to build a company that will employ our local people, help our fellow farmers and ranchers, and change the way food is processed and delivered."

"My plan is to open two separate companies. The meat division will be called Meadors & Scott Fine Meats, and the other company will be called Grandma's Meat Pies."

We all looked at Grandma when Andy had finished speaking. She looked like she was going to faint. This part of the plan was new to all of us. Andy explained that we could make thousands of pies and freeze them. That way they could be shipped and stored to be used, at a future time.

Grandma was still overwhelmed by the whole thing. Andy assured her that the company would employ lots of the women from town to help make all this happen.

Grandpa was trying to believe that this plan was possible, but it was just too much to absorb over a lunch table. Andy could see our dismay and told Grandpa that his job was to keep doing exactly what we were doing each day. He said that he would get the financing, supervise the construction, order the equipment, obtain contracts, and do the legal documents.

This was what Grandpa needed to hear. There was no way he and I could do more than what we were doing. There were many more things discussed, and Andy was very convincing. I truly believe that by the time he and his family left, he had us all believing that it was going to happen.

Over the next few weeks, our lives changed daily. There were trucks and workman coming early in the morning and not leaving until late at night. There were always construction noises and large pieces of equipment moving dirt. Wooden forms were created and trucks came full of cement that was poured for the new buildings.

Mr. Scott had a crew that was building holding pens for the livestock that would be coming to our future plant. The power company had sent out some men to bring extra power lines to our farm. I do not think that our town had ever had this much activity at any time in recent history.

Rarely did a day go by without a visit from Andy. He was always carrying papers and would talk to Grandpa for hours at a time. Before he would leave, he inspected every aspect of what was happening on our property. There was no way the plan would have ever come to pass without his constant direction.

One day a well drilling rig showed up and told Grandpa that they were here to dig us a very large well which would supply water for every activity in the new company. They had a map showing the site which surprisingly was nowhere near our spring. Grandpa helped them find the exact spot the water geologist had marked on the map. By the end of the day, they had already erected the drilling tower and told

Grandpa that they would start drilling the next morning.

At the end of just one month the foundations were all poured, and we could now see the footprint of each new building. Another crew had arrived and they installed septic tanks and a leech line for all the new bathrooms and showers for the workers.

Every three or four weeks, a representative of the government came by and took pictures and investigated the progress being made on the project. His name was Bill Wilson and on the latest visit, he had brought along another man. He introduced him to Grandpa and told him that Jack Renyolds was the individual who helped the military get vehicles of almost any kind, when they needed them.

Mr. Wilson said that Jack was here to see what the new company was going to need for transportation. He pulled out a list he had received from Andy and asked Grandpa to see if there was anything else that might be needed. I watched Grandpa as he looked over the list. He was speechless for quite some time.

Then he read the list out loud. Four refrigerated trucks, three pickup trucks, two trucks for hauling livestock, a flatbed truck for hauling hay, a large dump truck, a water truck, and two trucks with milk tanks.

The next list was for other equipment. It read, two forklifts, a road grader, a large tractor with a feed wagon, a second tractor with a frontend loader. When he finished reading the list, all he could say was, "I think that will do it except for one thing. We are going to need a much bigger fuel tank to fill up all this equipment." Jack said, "Consider it done".

Andy had asked for everything imaginable, and I do not think that he or Grandpa ever thought it would all be approved. But it was approved, and over the next few weeks equipment started showing up by the truckload.

Mr. Wilson had met with Grandpa and Andy to go over all the equipment for the new buildings. The list was very long but just like the truck list, it was approved. We would be getting a large room in each building that was refrigerated. There would also be a very large walk-in freezer room for each building. All the tables and work areas were to be constructed out of stainless steel.

Each building would have a large hot water tank with a long hose and a sprayer attachment so that every part of each building could be kept clean.

The main building where the meat was processed would have an overhead trolly system that would move large pieces of meat from one area to another and even into the refrigerated room.

In the smallest building which would house the pie making division, there were three large commercial ovens, huge mixing stations and several large work areas where assembling of the pies would take place. It had its own freezer room and refrigerated room.

By the fourth month, all the construction of the buildings was finished. The freezer and the other refrigeration equipment had arrived and was installed in each building. The lights had been installed. Restrooms were operating. The ground around the buildings was being leveled and graveled. The new well was working and sending water to a very large pressure tank which supplied water to the whole company.

Sitting in the parking areas was the entire list of vehicles we had requested. It appeared that we were going to have every part completed in just under five months from when we started.

All we needed now was some inside things like tables,

chairs, kitchen utensils, drinking fountains, and lockers for the workers. Everyday some of these were arriving.

Andy started interviews for the company. There were many of the men and boys coming back from the war with life changing injuries. Some lost an arm or a leg. These folks needed to work and help their families. Andy and Grandpa made the decision, that any of those veterans who wanted to work would have a place in our company.

The next step was to train the new workers. For that stage, we had to shut down some of our local deliveries for two weeks so that we could concentrate on training.

It did not take long. The workers were eager to learn and be part of something local. In the next six months, Grandma ended up hiring over forty local women for her pie business. Andy and Grandpa had twenty-three veterans and thirty-five men from town, working in the meat processing.

We were meeting our quota to the government and still using our forty percent to supply many communities around the area. The local ranchers and farmers were getting good money for their animals, and we even supplied the truck to pick them up and bring them to our plant.

Andy kept all the payroll records and paid our operating expenses. Stephanie Scott, Maggie's mother, had taken a very active role in helping Grandma.

Grandpa had been assigned a job of supervising. This was easier for him physically. I had much more freedom and was back to attending school during the week.

The Scotts and the Meadors had lunch together every Sunday after church. It had become a practice out of necessity. There were so many things for Grandpa and Andy to talk about, and Sunday seemed the only time they were able to be alone. Our plant was closed on Sunday so that the workers could attend church and have a day off. The other six days we were going strong.

The government agents visited us at least once every four weeks. Because of the demand for food to feed the troops, the government wanted us to run seven days a week. Grandpa had continued to tell them Sunday was the Lord's Day. So far, they had gone along with us, but we knew it could change in the future.

The first trouble came when Ed Evans showed up to ask for a job. Grandpa happened to be in the office when he came to the plant.

They went into Grandpa's office where they were alone. Grandpa asked him what job he thought that he could do well. Ed thought for a while and then said that he did not know but he needed a job.

Grandpa told him that the main job of the plant was to produce food for the troops and that the government was always checking on all aspects of the business.

Then he got personal and said, "Ed, I know you have a problem with the bottle and there are a lot of folks around town who have had run-ins with you. I am not sure this is a place for you. I am sorry but we have a very happy crew, and I am afraid that I cannot take a chance on giving you a job." Ed stood up and said "Oh, so now a poor dirt farmer is big time and can't help out one of the town's people."

Grandpa just said that he was sorry. Ed walked out and slammed the door shut as he left the office. Unfortunately, that would not be the last of Ed Evans.

## CHAPTER FOURTEEN

Life after the war was very different. For a while, people did not know how to act. Slowly at first some of the items that were purchased with ration books, started to appear in stores. Gasoline became available. People started to do some traveling. The major auto manufacturers started making cars again. The government started selling surplus equipment. Everything from trucks and cars to agricultural tractors and even airplanes.

Businesses and farmers took advantage of these sales and equipped their ranches and farms with machines that would replace less productive equipment.

There was a postwar boom. The government hired people to build roads and builders were encouraged to build houses. The government started a program where veterans could qualify to buy these new homes.

Our company kept doing what we did. We made meat pies and processed beef, hogs, and sheep. Maggie and I continued our schooling. We no longer had to work at the plant on weekends. Our Sunday dinners continued but became less regular.

On a cold morning in 1947, I went out to do a few chores and check on Ogi. He was all curled up in some hay. He looked like he was sound asleep. When I touched him, he did not move. Next to where he lay, was a stick. He must have had it there for our morning play time.

I picked him up and the tears fell from my eyes. Ogi had been Oganalii (my friend) as long as I could remember. He listened when I cried and told him my sorrows. He would place his head on my lap and roll his eyes up and look at my face. I swear he understood the things that I told him.

He was there when I learned to play baseball. He heard

the story of me being an orphan. He waited every day, by the house for me to come from school and play with him. My heart was broken. Deep inside I knew he was getting older and becoming frailer, but my heart was not ready for him to go.

I do not know how long I had been sitting there holding him, when Grandpa came in and understood what had happened. He sat down in the hay and put his arms around the two of us. I think that I saw a tear or two fall from his eyes. We sat there for a long time before Grandpa said, “Let’s find a nice place by the lake for Oganalii.”

We loaded Ogi in one of our trucks and drove to the lake. Grandpa let me pick the spot. We dug a deep hole and laid him in the bottom. I took off the shirt I was wearing and put it over him. We covered him up and put some flat rocks that we had found along the lake edge.

Grandpa said he would have one of the men make a nice wooden marker for the grave. We drove home and neither of us said a word. When we went inside Grandma looked at us and there was a question on her face. Grandpa just shook his head no. I went to my room and lay on my bed until I fell asleep.

When I got up it was obvious that Grandma knew about Ogi. She came over to me and hugged me with all her strength. I hugged her back. She never mentioned Ogi again.

When I saw Maggie at school the next day, I tried to tell her about Ogi without crying. But it was not to be. She listened as I told her the story then she put her arm around me and lay her head on my shoulder.

For many weeks after Ogi died, coming home after school was just empty. For most of my fourteen years of life, he had been there to greet me. Now, he was gone. Every time that I saw one of his sticks, my heart would ache.

I quit going to the lake. I did not throw baseballs. I went around the yard and gathered up all his sticks and burned them up. I piled bales of hay on the spot where I had found him so that I could not see where he had died. I tried to block his memory out of my mind. It did not work. My "brother" was gone.

True to his word, Grandpa had a marker made for Ogi's grave. It was simple it said Oganalii, donadagohvi. In Cherokee it said, "my friend, we will meet each other again".

Over the next few months, the pain in my heart slowly became bearable. It was a long time before I ever returned to the lake.

One afternoon Grandpa asked me if I would like to get a new dog. I told him that I would never own another dog. Grandpa looked at me for a very long time before he spoke. "Gule, the Creator made each of us in a special way. He gave us eyes to see His wonders. A waterfall, lightning in a summer storm, the first falling snowflakes of winter, and rainbows. But He also gave you a heart that can see. Sometimes, we are unable to see these special things because we are too busy being sad."

"Ogi was your brother. The Creator gave him to you for a time in your life when you needed him. And when you did not need him any longer, he was taken. The Creator never makes a mistake. Ogi was no longer needed to be your brother because the Creator gave you a sister. Her name is Maggie, and she has been there for you since you were six years old. It is time that you see your sister with your heart. I believe when you do that you will find a beauty there that you need in your life."

When he finished, he put his arm around me, and we walked back toward the house. My Grandpa was not a man of many words. But when he spoke, his words were full of meaning and wisdom.

For the next several days, I pondered his words to me. At times they seemed to make sense and then at other times, it was confusing. He was right when he said that Maggie had always been there for me. I just did not understand how she would replace Ogi.

At the next Sunday dinner, Andy told us that he was taking his family on a month-long trip to see the national parks. We had learned a little about them in school. Beyond what the teacher told us, I knew nothing. It sounded like it might be fun. Maggie seemed excited.

When dinner was over, I asked Maggie if she wanted to walk down to the lake. She said yes and we left as soon as we had helped clear off the table. It was a beautiful day, and the temperature was just right.

When we got to the lake, I grabbed her hand and took her over to Ogi's grave. I pointed to the marker and told her that one of the men at work had made the marker. She touched the marker and ran her finger over the surface, tracing each letter.

"Gule, what do the words say?" I told her that Oganalii means "my friend". The other words were like saying goodbye. However, in Cherokee there is not a word for goodbye, so we say donadagohvi, which means "we will see each other again".

She was silent for a moment and then said, "I think those are perfect words for someone you love." We stood there facing each other for a minute. I do not know what caused me to do it, but I reached up and smooth her hair. I looked into her green eyes, and my heart began to see for the first time. Grandpa was right.

Her next act was not expected. Maggie leaned into me

and kissed me right on the lips. I pulled away for a second and then I kissed her. She shoved me away and said, "Tommy Meadors, do you know how long I have been waiting for you to kiss me?"

I do not think that I had a correct answer to her question, I just smiled and kissed her again. This time it was a long, sweet innocent expression of my feelings. My heart was seeing more with each moment. Grandpa was right.

We held hands all the way back to the house. I felt all warm inside and I liked the feeling. Before the Scotts left, Andy told us that they were leaving in the morning. I did not realize that it would be that soon. A month was a long time. We wished them a wonderful trip. Maggie came over to me and whispered in my ear, donadagohvi. Then she kissed my cheek.

I wondered what everyone thought of that, but I was afraid to look. As they drove away, I waved until they were out of sight. In all the times that I had watched them drive away from our house. This was the first time that I was sad. Grandpa had been right.

The rest of the summer was busy but not out of control busy. Most of our employees had been with us for many years. They all knew their jobs and did them well. The truth be known, the plant would probably run without Andy or Grandpa being around all the time.

A week before the Scotts left for their vacation, Andy and Grandpa had a daylong meeting. Later he told us that they had decided not to try and grow the plant and make our business larger. They both felt that it was a good size and easily managed.

That was good news to Grandma and me. We wanted Grandpa to be able to take some time off and relax. During the war years, he had worked nonstop. He deserved the time for himself. Grandma felt the same. She once told me that she saw meat pies in her sleep.

The next evening after dinner, Grandma said, "So Roy, if you had some time off, what would you like to do?" Typical Grandpa, he had to think for a while. "Ida, I think that I would like to go to Tahlequah and show you and Gule things that are part of the Cherokee history." Grandma said, "If that would make you happy then I think we need to plan a trip."

Over the next week or so, we talked about was our first vacation. Because, I had never been outside our little town, this was an adventure for me. I now had two things that excited me. The return of the Scotts and our vacation. It would be many years before we would take that trip.

I spent most of my days driving a tractor around the plant and picking up garbage or hoeing a weed here and there. We were proud of how our business looked when customers drove up to our buildings.

One afternoon, I was dumping the trailer that I pulled around with the tractor, and I saw a car driving unusually slow down our driveway. I watched as it stopped and a man who I did not recognize stepped out and slowly looked all around.

Grandpa had always taught me to be friendly to the customers, so I walked over and asked him if he needed help. He gave me a big smile and said his name was Rex Walker. I shook his hand and told him I was Tommy Meadors.

That seemed to be of interest to him. He asked me if I was related to Roy Meadors and I told him that Roy was my grandfather. He looked over at my tractor that was pulling our garbage trailer. "So, you are the grandson of the owner, and you are collecting garbage."

I just laughed and told him that around here, we all do jobs which make our business look good for our customers. He smiled again and asked, "Do you think you could take a little time off from your garbage collecting and show me around?"

I told him that I could do that and then asked him what he would like to see. I will never forget his response, "Mr. Meadors, I want to see it all." I had never had anyone call me Mr. Meadors and it sounded very foreign.

I said, "Well since we are outside, you might want to see our holding pens and feed storage." That is where we started. He was not kidding when he said he wanted to see everything. When I would sort of rush from place to place so we could get finished with the outside tour, he would slow me down.

I never saw him take any notes, but later in the day, he would quote things that we had talked about in the morning. He asked more questions than anyone that I had ever showed around our plant. I finally asked him if he would rather talk to my grandpa and he told me no, that he was enjoying our visit.

I spent several hours with him and answered what seemed to be thousands of questions. When it appeared that his interest had been satisfied, he said "I am very impressed with your knowledge of your family business. How old are you?" I told him I would be sixteen on my next birthday.

He thought for a moment and said, "Tommy , I think it is about time that I meet your grandfather. Do you think that can be arranged?" I said that we could go see if he was in his office.

When we got to Grandpa's office, the door was open, and he was reading some reports about our monthly production.

I stepped in and told him that we had a guest named Rex Walker and he would like to meet the owner. They shook hands and Grandpa asked him " How can I be of service?"

Rex said that he represented a group three of Texas cattle ranchers, and they were looking for a successful meat processing business to purchase. He went on to say that in their search for a business that our company had come up several times. He finished by telling Grandpa he had never met a fifteen-year-old boy who knew and understood every aspect of the family business. He told Grandpa, "You got yourself a winner of a grandson."

Grandpa smiled and told Rex that I was his pride and joy. I was getting embarrassed by all this talk and asked If I could get back to my chores. I said thank you to Rex. He stood and shook my hand. "I hope we will be seeing more of each other Tommy."

When the plant closed at five o'clock, they were still talking. I went to the house to clean up for dinner. When I walked into the kitchen to see what Grandma was cooking, Rex and Grandpa were sitting and watching Grandma finish preparing the dinner. This was very unusual. We had never had anyone who toured the plant, come into our home at dinner time.

We sat down and Grandpa prayed as he always did. I sneaked at look at Rex and his eyes were closed and his hands were folded I was beginning to like this man.

During dinner Rex never asked Grandpa any questions. His focus was on Grandma, and he wanted to know everything about her part of the business. This was very unusual for Grandma she was used to being a small part of our main plant. Rex was interested in how she got her recipes and did she have any ideas for more products.

Grandma was very nervous at first, but Rex made her

comfortable with his soft Texas accent and gentle words. He made her contribution to our company seem important and praised her for her part in making Meadors & Scott a success.

When dinner was over Grandpa asked if Rex would like to sit on the porch for a while. He seemed very happy to have the invite. Grandpa sort of took the lead and wanted to ask Rex a few questions. He started out with simple questions and worked his way into the things which mattered to him. "Rex, you said you had two partners, and that you wanted to buy a company that could process your own cattle. That would be a big investment.

How many cattle do you own?" Rex responded, "Well Roy it is hard to keep up with the exact number. I would say somewhere over eighty thousand head. I think if Grandpa had not been sitting down, that he would have fallen over. This man sitting on our porch owned more cattle than our entire county combined. He was sitting here talking to my grandfather with respect.

Grandpa finally composed himself enough to say "Rex, that is a pretty big field to harvest." Rex laughed and slapped Grandpa on the knee. "Roy, I could not have said it better myself."

Rex finally said he should be going and thanked Grandpa and me for spending the day with him. But before he left, he went inside and thanked Grandma for a wonderful dinner and evening with our family.

As he left, he told us that he would be in touch and for us to think about any question we might want to ask on his next visit. We did not even know there was going to be a next visit.

As he drove out of sight Grandpa said, "Gule, that is a good man. It will be interesting to see what he has to say on his next visit."

During the next few days our main topic at the dinner table was Rex Walker and his plan. We were anxious for the Scotts to return from their vacation so that we could talk about our visit with him.

I did not know whether selling our company was a good idea or a bad one. There were so many things which would happen with a sale. Grandpa and I did not talk about it. It was easy to see that he would be glad when Andy was back and they were able to discuss many things.

Grandma was her usual self. Whatever happened was fine with her. She did not have an emotional attachment to her pie company. If Grandpa and I were happy, then she was happy as well. If someone gave Grandma a million dollars, she would have no idea what to do with it. Her lifestyle was simple, and she was very easy to please. Grandpa and I were blessed that she was part of our lives.

The Scotts arrived back in town on a Saturday evening. Andy called Grandpa and they talked for a short time. Grandpa invited them for Sunday dinner.

That dinner would prove to be the most important Sunday that our families ever spent together. Grandpa prayed before dinner and asked God to give us wisdom in our decisions. While we ate, He told the Scotts all about the visit from Mr. Walker. Andy only listened as Grandpa went on and on with every detail that he could remember.

When he finished Andy said, "Well Roy, you had a more eventful adventure than we did on our whole vacation." We finished eating but no one wanted to leave the table. There were to many issues to discuss. After three more hours of talking, Grandma offered everyone some cobbler she had made. We ate that and talked even longer.

The final decision was that each family would talk with

their members, and we would meet again at the office the next day.

On Monday we all met in the main office after the workers fired up the plant. Andy came ready with a list of questions that he had written. Maggie and I had not had any time to ourselves on Sunday.

I was looking for a reason we could skip the meeting. When Andy started to speak, I asked if Maggie and I could be excused because this should be a meeting for the adults. Maggie's mom smiled and nodded to Andy who agreed. Maggie and I almost ran out of the room.

When we got back to the house and we were alone she hugged me and said, "I missed you, Tommy Meadors." I said, "I missed you, Maggie Scott." We had a warm kiss and then I told her we should make a lunch and go to the lake.

She set about making the lunch and I got one of the company trucks and loaded an old wooden table and a bench in the back of the truck. I got an old tablecloth out of Grandma's rag box and off we went to the lake. When we arrived, we found a nice flat spot under one of the oak trees. I set up the table and the bench. Maggie placed the old tablecloth on the table and spread out the lunch. It was going to be our first private picnic.

The thoughts of selling our company were taking the back seat to my thoughts about this wonderful girl sitting with me by our lake. We ate the lunch, and she told me all about the vacation trip. It was hard for me to picture the things she was describing because I had never been outside our county. I did my best to listen and comment. The subject eventually got around to Rex Walker's offer to buy our company.

Maggie wanted to know how I felt about that possibility. I told her that I wanted what was best for Grandma and

Grandpa. It was important to me that Andy was also satisfied with the terms when we got to finally see an offer. I trusted him and I know he would take care of my family. It was so different to be thinking about life without meat and meat pies. Maggie and I had spent Saturdays and summers working at the plant for many years. It was hard to imagine it not being part of our lives.

When we were ready to head home, we decided to leave the old table and bench by the lake. This was going to be our special place when we wanted to be alone. On the drive home Maggie sat right next to me in the truck and laid her head on my shoulder. Life could be made so special when it included someone like Maggie.

I guess we had been gone two or three hours. Our families had moved to the porch, but they were still talking about the possible sale. They greeted us and asked how things were at the lake. We told them that our special place now had an old table and bench. Andy said, "Are you planning on moving down there?" That got a laugh out of everyone.

Grandpa spoke up and said that there was not much more we could talk about until we could see the terms that Rex wanted in a sale contract. Andy agreed with Grandpa and said let's make sure we have the things that are important to us on our list.

He started reading from his notes. The number one is that they must keep our employees. We do not want the name of the company changed. It is important that we do not sell one hundred percent of the company. We need some type of ownership that can be passed on to Maggie and Tommy. If they want to buy the entire farm, the lake and some of the surrounding property will not be included in the sale.

It was apparent that there had been some heavy discussions, to come up with these ideas. Maggie and I were very impressed with our parents planning.

Over the next couple of weeks both Andy and Grandpa received phone calls from Mr. Walker. It was very clear that they were still interested in buying our company. Rex told them both that he and his partners were planning a trip to our farm within thirty days. He said they would stay as long as it took to get a final agreement together.

Suddenly, a possible sale was turning into a probable sale. Andy and Grandpa had many meetings over the next few days. Their list of demands was slowly being defined in the proper form. There was not much more that could be done before seeing the offer from Rex and his partners. Grandpa had always told me that there is a reason that things happen. One of those unexpected things was about to show itself.

One evening, Mr. Amos showed up at our door. He said that he was there to talk some business. Grandpa assumed it was about wanting to farm the field that he grew corn in last year. That was not at all what he wanted to discuss.

When Grandpa realized it was something more important, he suggested that they go out on the porch. Mr. Amos got right to the point. "Roy, you and I have known each other our whole lives. You have always been a good friend and neighbor. I would appreciate it if you did not tell anyone what I am going to say. I am getting older, and farming is no longer what I want to do with my remaining years."

"My wife and I had a talk with our kids. None of them are interested in anything to do with farming. The bottom line is that we can see how well you are doing with your plant, and we want to sell our farm to you. We would like to retire closer to where are kids are living. I do not want to sell a piece ground that I have poured my sweat and tears into for more than 50 years. But, if I do sell, I want it to be to someone who will appreciate my farm. You are my first choice Roy."

Grandpa told us later that he was stunned. He had never believed that the Amos farm would be for sale. But Grandpa did not believe in coincidences. He felt this was happening for a reason. We might be selling our land and now we had the opportunity to pick up the land that was next to ours and would connect to our lake property that we wanted to keep.

Grandpa was very kind to Mr. Amos and did not question him about anything beyond what Mr. Amos had told him. Grandpa told him that there were some things that might change our plans and that they were coming up soon. He asked Mr. Amos to write up a list of things that would be included in the sale and all the sale terms. When he was ready, Grandpa offered to sit down with him and see if there was a way to make it all work.

Mr. Amos thanked him several times and told Grandpa he would get it ready very soon. Grandpa came back into the house after he left. Grandma could see that something important had taken place on the porch. Grandpa just looked at her and said, "Mr. Amos wants to sell us his farm." Grandma took off her apron and fell into a chair. "Roy, I do not think that I can handle any more news. Why is this all happening at the same time?"

Grandpa just looked at her and said, "Some things are just meant to be Ida." So now we had to consider selling our company and farm to Rex and purchasing the Amos farm. Grandpa said a few days later, that in all his life, he had never been in such a position.

At first Grandpa kept the offer to buy the Amos property

to himself. He told us that he did not want to add extra things for Andy to consider as we looked at the soon to have offer for our company.

A few days before Rex and his partners were to arrive, Grandpa got a phone call. It was from Rex. He told Grandpa that they were coming in five days. He said that he was bringing his accountant and attorney in case there were any legal or tax issues that needed to be finalized.

To Grandpa's surprise Rex told him that the first day that they would be visiting, they would not need to meet with him or Andy. He requested that Maggie and I be present to give them a tour of all the plant and property. Grandpa thought that was odd but told Rex he would arrange for that to happen. At dinner that night, Grandpa told Grandma and me what Rex wanted. He suggested that I call Maggie and make sure she was here that day.

The big day arrived. Maggie had come to the farm early and we all had breakfast together. It was still confusing why we were giving the tour and not Andy and Grandpa. But we were going to have that answer and a lot more by the end of this week.

Around nine o'clock, we could see a car approaching our farm. When it stopped by the house Rex and four other men got out. Andy and Grandpa greeted them, and everyone shook hands. Rex saw Maggie and me standing on the porch. He yelled, "Hey come on out here and meet these guys." We were given a warm introduction and then Rex said, "We are wasting time, are you and Maggie ready to give us the tour?"

He was a man who did not waste small talk. I liked him. Rex wanted a tour of the farm and then the meat plant, followed by a visit to the pie division. Grandma told everyone that she would have lunch ready by noon.

Our company had purchased a surplus twenty passenger bus after the war, and I suggested that it might be easier to get us around by using that bus. Rex agreed and we started our tour. I drove around the perimeter of the farm and stopped at the lake. We all got out and it was easy to see that they liked the lake.

That was not good news to either Maggie or me. Rex wanted to know about the old table and bench. Maggie told him it was our special place. He just smiled.

Someone noticed the grave marker and asked about it. We all walked over to it, and I explained the story of my dog Ogi. They asked about the words on the marker, and I explained to them the Cherokee meaning. Rex asked how it was that I understood Cherokee. I explained to them that Grandpa was part Cherokee and that we used a lot of Cherokee words in our home.

When they were ready, we went to the holding pens and the feed storage. The group did not ask many questions, but they did a lot of looking. It was almost noon, and I suggested that we head to the house or Grandma would be upset if we were late.

When we arrived, Grandpa showed the men where to wash up and then he asked me how the tour was going. I told him they liked the lake but did not comment on anything else that I had showed them.

During the feast of a lunch Grandma had prepared, Rex kept the conversation going on almost anything that did not have to do with the purchase. He must have had a plan because it was too obvious what he was doing.

When everyone was full and had thanked Grandma for the lunch, Rex said that we could now tour the meat plant. Over the next three hours we went over every machine and every space in the plant. They wanted to know the function of the machines. Could the spaces be enlarged? What was an average day's production? How many animals could be held in cold storage? The questions were endless. Rex was not taking any notes, but his partners were constantly writing something.

When we finished with that part of the tour, Rex asked the men if they had any more questions. There were a few which I answered. When there were no more questions, Rex looked at Maggie and said, "Ok girl your turn."

Maggie smiled and we walked over to the pie building. Maggie started the tour and as the question were asked, she did a great job of explaining every part of how that company operated. Rex hardly said a word. He just let Maggie run the whole tour. They asked her the expected questions. What is the daily production? Can the system be expanded? How many kinds of meat pies do we make? I thought Maggie did a great job. Rex must have thought so too, because when she was finished, he gave her a big smile and a thumbs up.

I asked them if they wanted to see anymore and Rex answered for everyone. "No Tommy, that is all we needed today. Tell Andy and your grandfather that we will meet with them at nine o'clock tomorrow in their office."

That was it. They walked back to their car and left. Maggie and I went to the house and spent the next hour telling Grandma and Grandpa everything that we could remember about our day with our visitors. It was obvious that Grandpa was confused as to why he and Andy were not included in the tour. The answer to his question would be coming very soon.

The next morning at nine sharp, Rex and company showed up at the office. Grandpa had a larger table brought in that morning so that there was seating and table space for everyone. Andy started the meeting and asked if there were any questions that needed to be answered before Rex and his people made a presentation. Rex said they were ready to move forward with a proposal.

He told us that the lawyer would read the proposal. Then the accountant would make some comments and that he would do the follow up.

The lawyer, Mr. Borden, opened his folder and started to Read. "Gentleman, my client Rex Walker and his partners are interested in acquiring Meadors and Scott in its entirety. That would be the property known as the Meadors Farm, the buildings, equipment, and transportation portion of Meadors & Scott.

The partnership would also want to acquire the entity known as Grandma's Meat Pies and the buildings and equipment that was currently in use. The partnership will prepare documents where all stocks in those companies are held by interested parties who will be members of a newly created holding corporation."

Before Andy or Grandpa could ask any questions Rex introduced the accountant, Mr. Roberts. His job was explaining all the numbers. "Gentlemen, thank you for the opportunity to present the proposal which we feel will take Meadors & Scott to a new level in the meat industry. Our goal is to be able to process our own beef through a company that we own.

To accomplish that we are willing to purchase the assets of Meadors & Scott which includes, the name, equipment, goodwill, and customer information. Our cash offer is $825,000. We would also want to purchase the property known, in these documents, as the Meadors Farm and all accessory buildings, wells and equipment for a cash sum of $190,000. It is also our goal to purchase the name, buildings customer list, goodwill and equipment for the company known as Grandma's Meat Pies. Our cash offer is the sum of $320,000.

It is the partnership's desire to employ Andy Scott and Roy Meadors for a period of one year. That time may be extended by mutual agreement between the parties. The salary for each of these individuals, shall be $75,000 per year and will also have health insurance supplied by the partnership. Meadors & Scott shall provide management duties and training of future company leadership."

When Mr. Roberts finished Rex stood up and began to speak. "I am going to make a few more comments and set some goals. Then we will turn things over to you both for any questions or items you may want to have clarified. Our partnership would like to have a place where we can have a regional feed lot and the processing for all our own cattle.

Your existing property and facilities offer us most of that opportunity. We are also interested in this relationship being long term. During the few days my partners have been able to see what I saw on my first trip here. Two wonderful families with lots of talent people. With that thought in mind, I want to include a few more things in our offer to purchase.

Tommy and Maggie know these businesses inside and out. I want them to feel that this new company can be there home for as long as they want to work here.

They have grown up in this business. Going to high school for another year or so, will do nothing toward the education that I want them to have. Therefore, I would like to pay for them to attend Texas A&M in the fall and take business courses as their main degree and the Tommy would also take animal husbandry. Maggie would take further study in dietary practices. I want Tommy to approach the feeding of our cattle as fine science. And Maggie will use her studies to make healthier and more nutritious products at Grandma's Pies.

I would want them to work each summer at our companies and upon graduation they will each receive five percent stock in their company, if they agree to work for the company for a minimum of five years."

Rex also said that the corporation would be looking for up to 1,000 additional acres of land. The closer to the Meadors farm, the better. The land would be used for feed lot and temporary holding of Texas cattle heading for the plant. He finished up, "Well gentlemen what do you think?"

Andy pointed to Grandpa, and he stood up. "Well, it seems you folks have thought a lot about this purchase. It will take me a little time to digest it all. I did want to clear up one thing. Our lake is not for sale. I guess the second thing is that Tommy and Maggie will need to make their own decision on going to college and working for your corporation if we sell the companies to you. I would not have any problem with working for you for one year if it would help train my replacement.

But that is just me. Andy will need to speak for himself. Grandma will do whatever I agree to, so you could figure on her for a year. I would need to talk to Andy about the pricing you have offered on the companies. The price for my farm is fair."

He sat down and Andy stood up. "Gentlemen I want thank you for your thoughtfulness on this matter and your offer. Like Roy, I cannot speak for my daughter. I agree with Roy, we need to talk alone about the pricing of the two companies. I also agree with the employment terms and pay." Rex stood up and said "How much time do you need to decide if you will sell? We can stay in town a couple more days if that is long enough for you to make your decision?" Andy looked at Grandpa and nodded. "Ok, then let's meet here in two days at nine o'clock." Rex asked when he might be able to talk with Tommy and Maggie. Andy said, "They are over at the house. I will go get them for you."

When Maggie and I returned, we were not prepared for what Rex had to say. He introduced us to everyone and said that it had been his pleasure to get to know both of us. He went on to tell us about the offer he made to Andy and Grandpa. Then he told us about going to college in the fall.

I held up my hand and he said, "What is on your mind Tommy?" I told him that we were not even finished with high school and that I did not know if I would even be qualified to attend a college. Rex told me that he and his partners have connections at Texas A&M, and we should not worry about the details.

I asked him about a few of the other details regarding our employment contract. Then it was Maggie's turn to ask questions. She was much more practical and asked things like, "Are we going to stay in a dorm?" "How do we get back and forth to school?" "How do we pay our monthly living expenses?"

Rex smiled and looked at his partners. "You see what I was talking about." Then he told Maggie that we each would have a little account which would be funded each month, and we could draw sums from that to purchase needed supplies.

When all our questions were answered, Rex said, "Thank you for your questions and for showing my partners why I have so much faith in both of you and your place in the future of the companies. You will be assigned a member of my staff that you can contact anytime day or night. Her name is Andrea Black. You will find that she is very capable and will take care of your needs and questions.

All I need from you is to say yes to my offer. And if your answer is yes, I will look forward to seeing you graduate in four years and take over the leadership of Meadors & Scott and Grandma's Meat Pies. We have great plans for the future of these companies and educating the next generation leaders of our company is part of our plan to make this venture a roaring success."

Both Maggie and I thanked them and said that we would like to speak with our families before making any decision. Rex had told us that the next meeting with Andy and Grandpa was in two days. We told him that we would be able to respond to them by that time.

## CHAPTER FIFTEEN

It goes without saying, a lot of time in those next two days was spent talking about the future of our families. We spent many hours discussing every aspect of the contracts for our parents and those for us and our schooling.

Andy thought the college plan was a huge opportunity and that if we signed the contract upon graduation that the five percent stock could be a wonderful thing for our futures.

Grandpa felt that helping transition the leadership by working for the new corporation was a sign that Rex and his partners were serious about a long-term relationship with our families, the community, and our workers.

Grandpa and Andy met alone and drafted a few more things that they wanted the sale to include. Maggie and I were encouraged to make a trip down to the lake and talk about our own futures with the company.

The day before the big meeting Grandma made us a big picnic lunch. We got some old clothes to swim in and drove down to our special table in our special place where we sat on our special bench and discussed our future.

Looking back over the years, that day with Maggie and all the things we talked about, was one of the most important days of our lives. Very few high school students are forced to make such long-term life changing decisions in just one day.

We did our best to take each issue and talk about it. Our first subject was going to college in the fall. We would be among the youngest students and would probably be the only two students who had not even graduated for high school.

It was overwhelming to us. Maggie pointed out that Rex

felt our practical experience in our family companies had prepared us in a way that most other students were not. She also mentioned his total confidence in us. I had to agree with her. Rex never wavered in his confidence that we would do fine in college.

We both felt that the support Rex offered us from Andrea Black would be very important in our success. We talked about being gone from our families. In one way it could be an adventure. In another, we would both miss our parents. Not working at the plant would also be hard at times. We had grown up in those companies, and the workers were our extended family.

We finished our lunch and decided that taking a swim would be a good break from all these heavy thoughts. We changed into our old clothes and raced to jump in the water. It was cool and refreshing. We played for almost an hour before laying in the sun to dry off.

Maggie lay with her head on my shoulder, and it was not long before we were both asleep. Maggie woke up first and found a feather on the ground. She used it to torment me by rubbing it on my face until I woke up trying to stop what I thought was crawling on me. She let out a loud laugh and kissed me. "Come on you lazy bird, we better finish talking so that we can make our decision." I did not know whether to kiss her or throw her in the lake. She was such a whimsical girl who I loved having as my friend.

We sat at the table and snacked on the remains of our lunch. We made a final choice to go to college in the fall. Both of us wanted to return after college and help continue the companies our families had started. We decided that we would ask our parents advice and if theirs matched ours, that we would tell Rex we were ok with his plan.

We sealed our decision with some kisses and headed

back to the house. On the way back, Maggie sat close to me and held my hand. It gave me a wonderful feeling. Her red hair and freckles created a delightful picture in my mind.

As we entered the house we saw everyone sitting at the table, which was covered with documents and notepads. Andy was the first to speak. "Well, I hope you two were able to make more progress on your decisions than we have been able to do." Everyone laughed. Maggie told them that with their support, we had decided to take up Rex on all his offers. I told them that we had some reservations, but Rex had shown so much faith in us that we were a little more confident in his plan for our future. The last thing that I said was "Maggie and I want to be a part of what the people we love the most, started in this very room."

By the time I finished speaking, Grandma had a tear or two in her eyes. Grandpa just smiled. Andy and Stephanie walked over to Maggie and hugged and kissed her. I went to Grandma and wiped her eyes with my fingers and hugged her tightly. Grandpa put his big hands around our shoulders and held us for a long time.

It was one of those moments that you never forget. It was quiet for a little while before Andy said, "We have a few things to talk about now that we have all our family members together."

Grandpa went first and told us that he and Andy had decided to take up Mr. Amos on his proposal to sell us his farm. We would use money from the sale of Meadors & Scott to purchase the farm. Rex wanted additional land for feeding and holding large number of cattle and having the Amos property would fill that need.

We would then have a farm that attached to the lake property which would not be sold to Rex. Our families would form a new corporation called A&L Land and we would give Meadors & Scott a long-term lease. This would keep the ownership of our land within the two families and give us all a place to build homes if we so desired in the future.

It appeared that the future was shaping up the way Rex had designed. Our parents would run the company for a minimum of one year. We would go to college and Rex, and his partners would become the new owners of our companies.

Grandma said that she had anticipated this being a long meeting and had a nice supper prepared for everyone. Maggie and I helped her set the table after Grandpa and Andy gathered up all the documents. Before we ate Grandpa prayed, "Dear Father, thank you for all your blessings to your children. Thank you for the opportunities before us. We pray that the decisions we are making will be guided by your plans for us. Help us to be good stewards with the money from the sale of our businesses. Amen." We had a wonderful meal and felt that we had discussed of all the things we should consider before signing any agreements the next day or so.

The day that changed our lives forever was here. It was nine o'clock. Everyone was gathered in the main office of Meadors & Scott. All six of our family members were there. Grandma, Stephanie, and Maggie from Grandma's Meat Pies. Grandpa, Andy and I from Meadors & Scott.

Rex had his two partners, along with the accountant and attorney. Rex spoke first and asked if we were ready with a decision. Andy said we were going to go forward with the sale of Meador & Scott, providing all the current employees would continue to work for the companies.

Stepanie said that Grandma's Meat Pies would accept the

purchase terms, providing all employees would still work for the company.

Grandpa said that he would accept the terms for the purchase of Meadors Farms if the contract was amended to not include the part of the farm containing the lake.

Maggie and I agreed to the terms of school and future employment. That left only one more subject to discuss. The Amos property.

Grandpa said," Rex, I guess there is just one more thing to mention. Andy and I have made an agreement to purchase our neighbor's property. That will increase our acreage and will attach to our lake property. The additional 1,000 acres you were looking for, will now be made possible.

Rex smiled and said "Roy, you are a better businessman than most folks would ever realize. I will have the attorney draw up a long-term lease on the property. This will allow us to move forward even faster with our long-term goals. Well done to you both."

"I would like to celebrate our new business relationship by having you come to our main ranch and see why we need your companies to expand our own."

He laid out travel plans and told us that they had anticipated our requests and had the changes made in the contracts to include our wishes. He said that it would only take a couple of hours to draw up the lease agreement and that if it worked for everyone concerned, have all the documents signed before he returned to Texas.

Everyone agreed and while the lease was being drafted Rex said he would like to tour the Amos property. I got the bus, and we all took the tour while Grandpa explained what the farm had been producing, the two wells on the property, and the buildings and equipment that could be included in the lease at a future time.

When Grandpa had finished educating Rex, we headed back to the office. The attorney had finished the lease, and he had placed piles of papers all over the table. He took each document and had the appropriate individuals sign in the places he had noted. It took more than an hour for everyone to sign and the lawyer to explain each document.

The accountant explained all the payments in detail. The Meadors Farms property would need to go through a local escrow process. The accountant showed us all the check which would be deposited in the account. These funds would go to Grandma and Grandpa as soon as the transaction was completed. There were two more checks. One was for Meadors & Scott and the other was for Grandma's Meat Pies.

The accountant placed one on the table in front of Andy and Grandpa and another in front of Grandma and Stephanie. That pretty much completed our business, and we no longer owned either company.

Rex told us that when he got back to his office, that he would arrange travel plans for all of us to visit the Texas operation. He said that while we were there, Andrea Black would finalize all the school plans for Maggie and me.

Rex expressed a desire to speak with the employees of both businesses before he left for Texas. We went to Grandma's Meat Pies first. Grandma made an announcement over the speaker that we had a guest who would like to meet with all the employees.

It took a few minutes to shut down the machines and then gather in the main room. Grandma introduced Rex and let him explain the new ownership. Rex did a very nice job of assuring everyone that their jobs were safe and that each of them was a valuable part of the company. He explained that even though he was the new owner, the leadership would remain in the hands of Grandma, Stephanie and Maggie. He thanked them for their part in making Grandma's meat Pies a success and said he looked forward to getting to know each of them. He said," There is one more thing. As the company goes forward each employee will receive an extra bonus based on the company profits for that year. This will be in addition to the yearly Christmas bonus."

I think that any reservations that they may have had over a sale of the company, were now put behind them. Everyone clapped and Rex waved goodbye as we headed over to the meat plant.

He made the same exact speech for the meat plant employees, and he got the same response. The plan was moving forward at lightning speed.

We all shook hands and Rex left with his group. I think our families were in shock. Until this moment all of this was a concept. Now it was a reality. When we went back to work at the plant, we were no longer owners, we were employees. Grandma was wondering about our house. Could we stay there or were we moving?

Grandpa held her hand and said that he had already talked with Rex and he had given us a year to decide what to do. Grandpa said that he felt we should remodel the Amos house and live there. There would be many more subjects to discuss, but for now, our futures were pretty much determined.

# CHAPTER SIXTEEN

Rex was a man of action. When he planned something, it was going to happen. Less than a week from the time he left our farm, we got a phone call from Andrea Black. She had a whole list of things to tell Grandpa.

We were to take a company vehicle and the six of us were to drive down to Texas one week from today. We would be staying at the main ranch which had several private cottages. Each of our families would have one for ourselves. When we arrived, there would be someone to greet us and show us to our quarters. She gave him the directions to the ranch and phone numbers to call if needed.

Grandpa had written everything down so he would be able to pass on the information to Andy. When he got off the phone, he told us everything that Andrea had said. Then he laughed and said, "This will be the first time that the companies have not had any of us here on a workday let alone an entire week. But we are taking orders from our boss, so I guess it will be ok." Grandma and I laughed along with him. How different our lives were becoming.

For the next few days, we worked at the plant and started packing for the trip. Andy and Grandpa had finalized then legal papers with Mr. Amos, and the escrow had been started. In a few weeks we would be the new owners of a very modern and prosperous farm. It was like some kind of dream.

I would see Maggie at work, and we would try to have lunch during the break time, but it did not always work out. I was looking forward to our trip to Texas so we could have a full week together and learn more about attending college.

The time passed quickly, and we were off to see Rex and

his ranch which was near San Angelo, Texas. It was about 550 miles from our home in Afton. We took a vote, and it was decided that we would leave early and arrive with one long day of driving.

Grandma had packed a picnic lunch for us to eat somewhere along the way. The time passed quickly because we were all talking about this big adventure.

After driving three hundred miles, we stopped for our lunch and Andy called Andrea with our location and expected arrival time in San Angelo.

The scenery was changing a little from what we had seen in the first few hundred miles. Everything to me was interesting because this was my first trip going anywhere away from our home. I guess that I really had no concept of the size of our country. Maggie and her family had done lots of traveling so this trip was not a big deal to any of them.

The afternoon passed and Andy figured that we should reach our destination before dark, without any problem. When he had called Andrea, she let him know that she would be waiting for us at the ranch.

It was almost six o'clock and we were seeing more cattle on the pastures that we were passing. We wondered if any of them belonged to Rex.

A little after seven o'clock, we came to a road that Andy was told would take us to the ranch headquarters. We turned down the road and came to a large arch that went clear over the road and the words painted on it said WALKER RANCH.

In the distance we could see lots of buildings. As we neared the buildings, we could see signs that directed the visitors to specific places. Andrea told Andy to come to the main office. We followed the signs and ended up in front of a very large building that had a sign which read Main Office.

Andy parked in front, and we got out, stretched a little and then headed up to the front door. Andrea must have seen us coming because she met us at the door and introduced herself to us. She looked younger than I had expected. In her boots, she was quite tall, with long brown hair and a big smile. I guessed her age to be twenty-five or so. I wondered how she was able to have such an important job at such a young age. Rex must be very good at picking talented people to work for him.

Andrea gave Grandpa a folder of papers that he and Andy were to look over before a ten am meeting the next morning. She gave Grandma a different file for her and Stephanie to look at before their eleven am meeting in the morning. She also had a folder for Maggie and me to look through before our one o'clock meeting the next day.

When she was done with that phase of our introduction, she showed us around the rest of the main office building, including a conference room where we would be meeting the next day. Next was a building that was smaller and just to the north of the main office.

Andrea said, " This is my favorite place. We call it the Chuck Wagon. It is a very quality buffet where all our employees eat for free. You should take advantage of that service anytime you are hungry. It is open Twenty-four hours each day because we have staff and employees working around the clock."

The next stop was medical clinic where an on-duty nurse could treat injuries that the ranch hands might get while working on the ranch. And there was a dentist office open five days a week. The last building contained a barbershop, leather shop, and a small store where one could purchase everyday items.

I was curious why the workers did not drive the 55 miles

to town for these services. When I asked Andrea, she told us that the ranch employed over 300 hundred workers and Rex wanted them taken care of on-site, so they did not have to make the long trip to town.

The other side of the street had three long buildings that we were told were bunk houses for the unmarried ranch hands. We did not go into that building but Andrea told us they were first class, with each employee having a private room. There were several small homes behind the bunkhouse buildings. She said that these were the housing for ranch mangers, and workers with families.

The next group of buildings were small cottages constructed in rows of six. There appeared to be three rows. Andrea told us they were the guest quarters. She stopped in front of the third and fourth cottages and told us that these were to be our accommodations, while we were guests of the ranch. Andrea asked if there was anything else she could do for us. We all said no, and she said that she would be at each of the meeting the next day. She gave each of us a business card and told us to contact her for any of our needs.

We unloaded our suitcases and explored what would be our quarters for the next few days. Everyone thought they were first class. Maggie came over to see how our place compared to theirs. She said they were both clean and looked very comfortable.

The parents decided that they were too tired to go to dinner, so Maggie and I walked over to the restaurant by ourselves. It was very nice inside and we were very happy to have some alone time. Maggie had brought our folder with her and since the restaurant was always open, we could take our time eating and looking through the papers.

There were maybe a dozen other people eating when we

started down the food line. I had never been to a buffet, and it was an amazing experience for me. There was a very long line of different food items to choose from, and it was overwhelming. I wanted to taste everything, but my plate was too small. I decided that the best approach was to eat a plateful and return for different things.

Maggie told me she had been to these places several times and she had learned the secret was to take a little of a lot of things and not a lot of anything. Her plate had many more items than mine, but in much smaller amounts. It soon dawned on me that there was a lot about life away from our farm that I needed to learn.

We had a feast and even two kinds of desert, before Maggie said that we should open the file and see what was inside. There were several papers some had Maggie's name on them, and a few had mine.

Maggie separated them into piles. The papers that did not have a specific name, appeared to be for both of us, so we started with those. One was a welcome letter from Rex. The second paper explained the leadership of Walker Farms. The third was an information sheet that covered many the business interests of Rex and his partners. The last was general information about the ranch itself. The number of acers of land, cattle, lakes, wells and crops produced were all listed. It was really mind boggling to comprehend the vastness of the ranch empire.

We decided to look at my papers next. They were sort of a restatement of our agreement. I was informed that my date to appear at the college was September 1, 1950. I would be met by Andrea Black. She would introduce me to another student who was the son of one of the ranch foremen.

He would show me around the campus and help me find

my classes for the first couple of weeks. His name was Darren. He would take me to see my school counselor who would help guide me along the path Rex had chosen for me to follow. The last paper said that my schedule of classes would be mailed to me in a couple of weeks.

Maggie had a group of papers that were about the same as mine. Her contact would be Janie, who also had a father who was one of the ranch foremen. She read that her list of classes would be mailed in about two weeks.

By the time we finished with these folders, it was getting late. We agreed to meet for breakfast in the morning and said good night at her cottage door.

Grandma and Grandpa were still awake when I came home. We talked for some time about our folder contents. Andy and Grandpa were going to meet for breakfast and go over their package. Grandma and Stephanie had plans to do the same. We were very tired and sleep came easily for all of us.

Everyone showed up for breakfast, but each person sat with their partner for the meeting. Maggie and I had already gone over our papers, so we had decided that when we were finished eating, we were going to go explore the ranch headquarters.

We walked to the tables where our parents were sitting and told them our plans. We left them alone, because they had a lot to look over before their own meetings.

As Maggie and I left the Chuck Wagon, we saw Andrea going into the main office building. We thought it would be nice to say good morning, so we headed her way. When we opened the door, she was talking with the receptionist at the main desk. She saw us come in and walked over to greet us.

"Well good morning you two. All ready for your meeting?"

We told her that we had prepared the night before. She smiled and said, "So what are you up to?" We told her we were going to do some exploring. She told us that there were several ranch vehicles parked behind the building and if we found one with the keys in it, that we were welcome to use it. We thanked her and she said that she would see us at one o'clock.

It did not take long to find something with a key in it. Maggie wanted me to drive, so we headed out on our adventure. We had seen most of the buildings near the main office, but there was a very large one a short distance away. As we approached the parking area, there was a sign saying RECREATIONAL FACILITY. Maggie wanted to stop and look inside.

As we entered there was a check-in desk and a young man sitting there doing some paperwork. He asked if he could help us and we told him we were visiting the ranch for some meetings. "Well, let me show you around." As he gave us the tour, we learned that the building housed a swimming pool, basketball court, fitness room, and showers.

He told us it was open twelve hours every day of the week. He was very friendly, and we thanked him for his time. We got back in the pickup truck we were driving and headed for some more buildings that were about one-half mile away.

As we approached there were several people going between the buildings carrying things and driving vehicles. It was a very busy place.

Next to several of the buildings were holding pens and corrals. We could see horses and cattle in all of them. Maggie pointed at a sign which said BLACK SMITH. We parked in front because it was a central place in the series of buildings.

The black smith shop had lots of equipment for working with metal, but the main activity seemed to be shoeing horses. There were three farriers inside and each was busy crafting shoes for the horse they were attending. A couple of them waved at us and we just waved back and kept moving toward the exit door.

The next building was a little smaller and had a sign reading VETRINARIAN. We did not go inside, but looking in the window it was as clean as a doctor's office. We could hear livestock noises from the back, so we assumed that they had some animal patients.

The next building was very large and smelled of leather. The sign on that building said TACK. We peeked into an open side door and there were what looked to be hundreds of saddles, bridles, saddle blankets, and even saddle bags. I had never seen that much tack in my entire life. Maggie was speechless.

We moved on to a smaller building which simply said BRANDS. No one was working there, and the door was locked so we just looked in the window. There were pictures on the wall of lots of branding irons and individual brands. It looked more like a museum than a business. But along one side were several racks of branding irons that obviously belonged to the ranch.

The more we saw the greater was our awe at the size and complexity of Walker Ranch. Our time was getting close for our meeting, so we headed back to headquarters.

We decided to get a light lunch before the meeting. While we were eating Maggie said, "I cannot believe that the owners of all that we have seen are interested in us being part of their empire." I had to agree. This ranch or business or enterprise or whatever you wanted to call it was beyond anything that either of us could have ever imagined. We finished our meal and crafted a couple of questions that we were going to ask. That was about all the time we had as it was almost one o'clock.

As we entered the main building, Grandma and Stephanie were leaving. Grandma was smiling and just game us a thumbs up. We went down to the conference room and knocked on the door.

To our surprise the door was opened by Rex himself. He had a huge smile on his face and said, "Hi you two, come on in." He pointed to some chairs, and we could see Andrea was already waiting for us. Having Rex at the meeting was a little wrinkle that we had not anticipated.

"So, I hear you have been checking out our little ranch." I guess word got around quickly with regards to strangers looking around. I was trying to find some words when Maggie spoke up. "Mr. Walker, Tommy and I have never seen anything of this size in our lives. It makes me feel small and we were wondering why you would want to have two small town kids working for you."

Rex thought for a minute and then said the words I will never forget. "Maggie, you and Tommy are exactly what I am looking to for the future of our ranch. You are both humble, and kind. You show respect for your elders. This ranch, seems overwhelming, yet you are not afraid of the responsibility. We have a method by which we operate. It is honesty, fairness, respect and hard work. We take care of those folks who take care of our interests. There is no single person who can run this ranch and our other businesses. It takes a team of hard-working and talented people.

We are interested in training you both to be two of those individuals when you finish with school. In the meantime, you will be working at the businesses that you know personally, and we will be teaching you our way of doing things."

He turned the meeting over to Andrea who gave us each some papers. We read them together. It was mostly things we could expect at school. They had arranged for each of us to have a roommate who was from the ranch. We would need to take a couple of tests which Andrea said were just a formality.

The guidance counselor that each of us was going to have, would be aware that our future was already decided and they were to only help guide us along that path. If we were to experience any problems with a class, we were told to contact Andrea, and she would find us a tutor to help with the class.

It seemed like they had thought of everything. I was beginning to wonder how many other kids, from the ranch, Rex had sent to college so they could someday help run his businesses.

Rex asked if we had any more questions. Maggie asked, "How do the children from the ranch attend school?" Rex told us, " It is a problem that we hope will be solved in the future. But for now, the ranch has some school buses, and we take the children to school and back each day. The drive is an hour each way and it makes the day long, so we are looking at alternatives."

That seemed to satisfy Maggie and since we had no more questions, Rex told us that he would see us in the morning for our big tour of the ranch. We walked back to our cottages and saw our parents sitting on the porch talking.

Andy asked about our meeting. I related all that I could remember and then asked about his meeting. Grandpa spoke up and said that it was mostly Rex explaining procedures, that he wanted to start at both our businesses.

Things like their accounting methods and inventory control. Nothing unusual or unworkable.

Stephanie said that their meeting was pretty much the same. Everyone was looking forward to the tour in the morning and our folks wanted to know what we had discovered on our own little venture that morning. Maggie filled them in about all the different aspects of the ranch that we had seen. They were as impressed as we had been.

Grandma wanted to take a nap and Andy and Grandpa wanted to do some more talking. Stephanie, Maggie and I got something to swim in and headed for the pool. The pressure of having the meetings behind us, seemed to make everyone more relaxed and now we would be able to enjoy the rest of our visit.

We had all been impressed by what we had seen thus far, but nothing could have prepared us for what we would see the next day. I slept like a baby and woke up ready to go. Grandma not only had a nap but had gone to bed early and did not wake up until after I did. Grandpa was his usual self. Able to function on whatever sleep came naturally to him. He was very excited to see more of the ranch operation.

We got ready and had set a time to meet the Scotts for breakfast. When they arrived, they seemed refreshed from their sleep as well. Everyone had a good appetite, and we were in good spirits. At the agreed time, a bus pulled up in front of the Chuck Wagon, and we climbed in to see both Rex and Andrea waiting for us.

Rex told the driver that we were ready, and we headed toward the buildings that Maggie and I had seen the day before. As we neared the complex, Rex started talking about this portion of the ranch headquarters.

There were at least a dozen buildings of various sizes. Most of them had a sign which stated the use. We parked in front of one of the larger buildings. Rex told us it was a quarantine facility for animals that were sick. Next to it was another vet clinic. This one was much smaller than the one we had already seen. We did not go in but, looking through the window, we could see cabinets with medicine and lots of vet tools. Attached to that building was a covered area that had pens and chutes where animals could be separated and doctored one at a time.

The next building had a tack area which was much like the one at the headquarters. Next door to that was a building labeled "Breeding". Rex explained that here, individual animals could be artificially breed. There were several freezers where semen from bulls and stallions was stored for use as the ranch tried to improve the bloodline of the animals they raised.

Rex had us walk to the next area and there was a building that stored everything that one could imagine. It was necessary for this purpose so that never ending trips to town were not required. As we walked through the building, Rex pointed out various supplies. I had never even seen a store with the number of products which were stored here. It was amazing.

We did not tour the rest of the buildings as a sign told All the activity within each of them. Rex wanted us to see more of the ranch. We boarded our bus and headed away from the buildings. For almost the first mile we drove down, both sides were covered with holding pens and thousands of cattle. Rex said these were holding pens where the cattle were feed a diet rich in grain and alfalfa. This was to "finish" them before they were sold.

The pens gave way to field after field of green grass. Some fields had cattle and others had none. Rex explained that the cattle were moved from field to field so that the grass could have a chance to regrow.

After about a thirty-minute drive, we came to a large lake. I was surrounded by large oak trees. There were cattle standing in the shallow water. They all turned to look at us. I saw hundreds of ducks sitting in the shade under the trees. Our little lake would have been lost in this huge body of water.

Rex told us that they had seventeen lakes on the ranch and everyone one looked about like this one. I was unable to process the amount of cattle we had seen, and the vast grazing lands.

We drove for another two hours before Rex said that he had arranged a special treat for us. We were nearing another lake, and we could see under the oak trees was a group of people who were preparing food. When we got out of the bus, Rex said they are preparing a real Texas barbeque for us. I think we were all hungry and the smell of the cooking meat only made it worse. One of the ranch hands told Rex it would be ready in a few more minutes.

There were some tables and chairs set up in the shade of a massive oak grove. The main food table had potato salad, buns, beans, tortillas, and some drinks.

As promised, several platters of meat were added to the table, and we were told to load up our plates. Maggie and I were motioned to start the line, so we took a plate and started down the line of food. I remembered what she had told me at the buffet, and I took a little of a lot of things.

I had ribs, steak, a hamburger patty, and a bun. I decided to not fill up on salad or beans. Maggie took some salad and steak. I was surprised that she chose only two items. She told me that she wanted to try one meat at a time so she would be able to enjoy the individual flavors. She was already sounding like a dietician.

I did not see what the others were getting until they came to the table. We all waited until Rex was seated. To our surprise he prayed before we ate. I saw the workers beginning to get their lunch and go to their own table next to the food area.

I can only say that I had never had food that tasted as good as this meal. I was already getting full when I finished my plate. I guess Rex could read my mind because he said, "Tommy, I think it would be wise for you to take another trip to the food table. A meal like this does not come around every day."

He was sure right. I had never had a meal like this one. Even Maggie had seconds. When we were finished, one of the workers approached our table with an apple pie. Not one of us had any room for pie.

Rex told us that we had only seen a small part of the ranch, but it was just more of the same. He said we could go back a different way just to change the scenery a little.

My stomach was full. My mind was calm and before I realized it, I had taken a little nap. When I woke up, Maggie smiled at me and said, "Oh the sleepy boy woke up."

I think this day had been one of the best days of my life. I was with my family and the Scotts. We had a wonderful time exploring a ranch owned by my new boss. Oh, and had a meal that was meant for only the rich. And I was able to share the entire experience with Maggie.

On the way back we stopped a couple of miles from the headquarters and Rex showed us a large facility that had stalls for about fifty horses and some pens with maybe a hundred head of cattle. He told us that they trained cutting horses in the arena and that the cattle we saw were for training purposes.

When we arrived back at the main ranch buildings, Rex thanked us for taking the tour and said that he was needed at a meeting in Dallas the next morning so he would be leaving. We thanked him for the tour and the wonderful lunch. He surprised us when he said donadagohvi.

Grandpa said the same back to him. We were told to meet with Andrea first thing in the morning. She would have a few more papers for each of us. Following that meeting we were free to go back home.

We had a light dinner, and our parents settled in for the evening. Maggie and I took a walk around the compound and enjoyed our time alone. The moon was out and there was plenty of light for our little walk. I was holding her hand and looked at this wonderful person next to me. She was no longer a child or a girl. She was developing into a woman.

At that moment in time, there was no place that I would have rather been. The air smelled of hay and cows. The gentle wind blew her hair away from her face and I was beyond content just looking at my best friend.

Maggie broke the silence and asked me, “Tommy, do you sometimes get a little overwhelmed, by all of this?” I was glad she asked me because I thought that I was the only one feeling a little insecure. I told her the truth, “Maggie if you were not doing this with me, I would not be going along with this plan that Rex has for us.”

She gave my hand a squeeze and said, “I feel the same way. Thank you for being honest with me.” That opened the door and we both shared our very personal feelings about everything that was happening. That conversation was one of the most important things that I had ever discussed with another person. We climbed up on a wooden corral fence and shared our inner most thoughts for another hour. We both started to fade a little, so we decided to call it a night.

I walked her to her cabin and kissed her good night. What a wonderful day. When I opened our cabin door Grandma and Grandpa were already asleep. I was barely in bed when I fell into a deep sleep. All the stress of everything seemed to melt away when Maggie and I talked. I was at peace.

The next morning, we all met for breakfast and met with Andrea. She gave us more folders and said that they were self-explanatory, so we did not need to talk about the contents. She said that we should go back and implement the record keeping and inventory control that Rex wanted us to make part of the companies. She said to expect a visit by Rex in the next thirty days.

We thanked her and loaded up for the trip home. Andy announced that we were going to enjoy the trip home so if we saw anything that looked interesting or fun, we were going to stop and check it out. He thought we should spend the night somewhere along the way and have an easy trip home the next day. Everyone agreed.

We stopped several times at roadside attractions and ended up only traveling 370 miles that day. The next day we did not see much reason to stop, and we were home in the early afternoon. Before the Scotts trip to the National Parks, Andy had purchased a new camera. He had brought it on our trip, and he must have taken hundreds of pictures. As Andy left for their own home, he promised to have them developed and share the pictures with us.

The next day, we were all back to our normal jobs. Grandma wanted to meet Stephanie and they would set up the new systems that Rex wanted. Grandpa and Andy did the same on their side of the company. Maggie and I did what we had done for most of our lives. It was business as usual.

## CHAPTER SEVENTEEN

Over the next few weeks, the two companies had implemented all the programs that Rex had requested. Andy, Grandpa and I had spent quite a bit of time over on the Amos property. We were trying to figure out a plan for future use of the property for ourselves and for the Walker ranch.

Andy had our lawyer look over the equipment list and other assets of the farm so we could determine a reasonable value of everything other than the ground. Because Grandpa was the one who was able to agree to the purchase with Mr. Amos, Andy felt that our family should have the right to live in the house. He was content to live with his family in town.

Grandpa now had the blessing to go ahead and fix up the home where we would be living. It seemed to be a burden lifted from Grandpa because he was much happier about losing our home where he and Grandma had spent so much of their lives.

Rex had called and was sending out a survey team to map the Amos property and to determine the best layout of the holding pens, pastures, and water distribution. The terms of the lease were that Rex would pay all the development expenses for the entire property less the costs of remodeling the house.

Rex wanted us to know that the survey team would be there before he was able to visit. Grandpa said that they could stay in the Amos house while they were mapping the ranch.

The time was passing quickly, and Maggie and I were spending time alone whenever possible. The time was approaching when we would be leaving for school, and life would become much different for us. Rex made his visit and was happy with how business was doing in the companies. He told us that when the survey crew finished on the Amos property, that they were going to work on an additional plan for the Meadors property.

That made sense because any expansion of the businesses would need a well thought out plan. That would prevent any conflict in the future for space and design.

When Maggie and I were making our final plans for travel to school, in ten days, our packages arrived from Andrea. Each folder contained our schedule of classes, the name of the teachers and our career counselor information.

It was getting real. We were leaving for school in a strange place, living with people we did not even know, and studying things we had never even thought of learning. Maggie was much calmer about all of this than I was being.

Without us knowing, Grandma had planned a going away party for us with the entire staff of both companies being invited. It was to take place after work on the coming Friday. Andy and Grandpa had promised to try and have a barbeque as good as the one we had in Texas. Grandma and Stephanie spent two days getting the salads and desserts ready.

When Friday came, Maggie and I still had to work, but we were excused a little early so we could change clothes and get ready. The workers set up tables and chairs in the main break room. Promptly at five o'clock the plants shut down and the party started.

Slowly people started entering the building. There were even some folks from town. When everyone was gathered Andy made a little speech about us going off to college. That was followed by Grandpa saying a prayer before we ate. Some of the workers from the meat plant had been working the barbeques most of the afternoon. When Grandpa

finished praying, they brought in huge platters of steak, ribs and of course hamburger patties.

Everyone formed a line to get their food. Maggie and I got to go first. Maybe it was because I was hungry or maybe it was just such a wonderful meal with so many folks that we had grown up knowing. I do not know which, but this was just as good as the meal Rex made for us In Texas. When everyone was done eating, Maggie and I stood up and each person got the opportunity to walk by and give us their best wishes. It was one of the nicest things that had ever happened to me.

Within an hour everyone had pretty much said goodbye and they had headed home. A couple of the men and a couple of the women were paid to stay after and help clean up.

Maggie and I were given permission to go to town and see a movie at our little movie theater. We had both worked all day and then sat through the dinner for us. We were tired but excited about the future.

We got through the movie without falling asleep and I took her home. I walked her up to the house and we kissed for a few minutes. That woke me up. Maggie finally suggested that I go home and get some sleep.

The plan was for our families to have our last Sunday meal together before Maggie and I left on Monday morning. That left all day Saturday without any responsibility.

Saturday morning, I called Maggie and asked her if she would like to spend the day at our lake. She said she would and could be here before lunch. I had a few things to do before she got here. I told Grandma that Maggie was coming. She said that she would make us a lunch. There was still lots of food from the dinner the night before, so she put together another feast for us.

Maggie got there about eleven and we headed to the lake with our lunch. It was still warm in August, so we were going to have lunch in the shade.

When we arrived at the lake, we moved our old table into the shade and set up the lunch that Grandma had put together. Maggie wanted to take a swim before we ate our lunch. We had a good time of jumping off the bank and trying to dunk each other. We wore ourselves out and just laid on a blanket before eating.

We talked about the trip on Monday morning. Grandpa had told me that Rex was going to allow us to take a company vehicle to Texas A&M. I had picked out one of the pickup trucks so that we had room for our school supplies and clothes.

I told Maggie that I would be picking her up at eight on Monday morning. We spent the next hour talking about the trip. By the time we had finalized our plan, we were both getting hungry. We sat at our special table and enjoyed another wonderful meal.

When we finished eating, Maggie said she wanted to lay down in the shade and maybe take a nap. That is what we did. Both of us dozed off for more than an hour. I woke up first and just raised up on my arm and watched her sleep.

Maggie was the best friend that I had ever had. She had found a place in my heart. I liked being around her. I hoped my future included Maggie.

The next day after church, the Scotts showed up for what would be the last Sunday dinner with Maggie and me for quite a while. We had leftovers from the Friday feast. Everyone thought it was still delicious.

We visited for a few hours. As the Scotts prepared to leave, Grandma hugged Maggie and said, "You take good care of my boy." And then she kissed Maggie on the cheek.

I was a little surprised by Grandma's affection to Maggie. Grandpa gave her a hug when Grandma finally let her go. They would not see her again before the Christmas break.

The next morning, I finished packing my things in the truck. Grandma had a nice breakfast ready and a picnic basket for our trip. We did not talk much because we knew that this would be the first time I had ever been away from our family.

When we walked out to the truck Grandma just grabbed me and began to cry. I felt like crying because she was crying. Grandpa came over and the three of us just stood there holding each other. Eventually Grandma wiped her face with her apron. She let me go and said, "I love you Gule."

That was all it took. My eyes began to water, and I was unable to speak. I was finally able to say, "I love you both." Then I turned and hurried off to the truck. It was only about fifteen minutes to Maggie's house. It took me that whole time to get my emotions under control.

When I stopped in front of her house, it was exactly five minutes before eight. The Scotts were standing in front of their house and Maggie's things were stacked up beside them.

Andy and I loaded her things in the truck, and we were ready to leave. He shook my hand and said, "Tommy, take good care of my girl." I told him that I would. I was doing ok until Stephanie gave me a hug goodbye and kissed my cheek. Just like at my home earlier that morning, we ended up in a group hug. Maggie and Stephanie were crying. I was doing my best to hold it together.

After a few more minutes we were able to get in the truck and start our journey on this new adventure.

# CHAPTER EIGHTTEEN

Maggie sat as close to me as she could in our truck. She did not speak for a few minutes. Just sat there snuggled up against me. Finally, she said, "Well here we go. Are you excited?" I told her that I was, but it had been hard to leave Grandma and Grandpa. All she was able to do was nod. She was trying hard to be strong.

After an hour or so of driving, we were both able to talk about the months ahead of us. It made the trip go faster. As the time was nearing midday, we were about to cross a river, and I suggested that we could stop there and eat the lunch Grandma had prepared. Maggie thought it was a good idea. We turned off the main highway and found a parking area under some oak trees right on the bank of the river. It was a perfect spot to have lunch.

I told Maggie that we needed to watch the clock so that when we got to school, we would have time to check in before closing time. We had a great lunch and a little break from the driving. Neither of us wanted to talk about leaving home. It was still very emotional. The school was our main topic.

We kept to our time schedule and were soon back on the road. When we were about an hour from the college, we stopped, and I called Andrea with our arrival update time. She told me that both our new roommates, Darren and Janie, would meet us at the check in desk.

It was all becoming more real with every mile. Maggie and I were both getting a little more focused as we arrived on campus. We followed the sing to the registration office. I parked and we walked up to the big doors for the entrance.

Darren and Janie must have seen a picture of us or had some way that they knew it was Maggie and me. They came over to us with a big smile and gave us a warm welcome.

With their help, the check in process was a piece of cake. Within a few minutes we were off to the girl's dorm where we helped Maggie get her things into her room. Darren did not want to stay long so I said goodbye to Maggie, and we left for the boy's dorm. Darren helped me with my things and then showed me where to park the truck.

I spent the evening talking with Darren. He had everything planned for a full campus tour in the morning. I made my bed and put my things away. It had been a long day, and I was tired. Darren was very nice. He said that he was going to the library for a couple of hours and that he would be quiet when he came in later.

The next thing I remember was Darren saying, "Tommy, it is time to rise and shine. We are going to meet the girls in less than one hour." I really needed to use the restroom and get a shower. Darren showed me where it was, and I made immediate us of the facilities.

We met the girls at the dining hall. Maggie looked all rested and ready to eat and then go on the tour. The breakfast was surprisingly good. Not like Grandma's but still good. When we finished, Janie took control, and we headed out for the tour. Both Maggie and I had brought our class schedule and a note pad. Janie had Maggie's and Darren had mine.

By lunch time, we had visited each one of the class locations. Maggie and I had taken notes so we would be able to find the classes on our own. No one seemed to be hungry yet, so Darren suggested that he show us the fitness center, pool, basketball courts, and other activity centers.

He saved the best for last. The very famous football stadium. It was huge and Darren told us it could hold 40,000 spectators. He told us that when there were home games, everyone on campus would attend the games. I think that the entire population of our hometown could fit in just a couple of rows of the seats.

Janie said attending a game in the stadium was like nothing she had ever experienced. We were convinced and made plans to attend the first game together with Janie and Darren.

When the tour was finished, Janie excused herself saying she had some things that needed attending. We thanked her and Darren for all their help. Before he left us, Darren said that we were to call Andrea in the morning.

For the first time in this whirlwind of activity, Maggie and I found that we were alone. She told me she liked Janie and her room was nice. I agreed with her. My room was nice and so was Darren.

We had seen most of the campus, so we decided that it might be fun to drive around town and see the sights. We spent the rest of the day and early evening checking out the area.

Because our meals were all paid by Rex, we had made the decision to eat as many meals as possible in the dining hall, rather than pay for a meal somewhere else. We had skipped lunch and were getting hungry, so we went back to the campus and had dinner.

There were lots of places on the campus, to sit and enjoy the scenery. We found one after dinner and cuddled up and just enjoyed the stars coming out. Our first day of school was in two more days. That left the next day for us to purchase books for our classes and make any final plans for our first official day of college.

I walked Maggie back to her dorm where we kissed good night. I got back to my room and Darren was not around. I went to bed and was reading my schedule of classes, when

I guess I fell asleep.

I met Maggie for breakfast. She had a list for the bookstore. I had one as well. We purchased everything on our lists and took them back to our dorms. We decided to see if we could find our first class for the next morning without using our maps. Maggie was able to locate hers, and I was not. Because we had different majors, the only classes that we would have together, were two business classes.

They were both in the afternoon. It was not shaping up to be a schedule that allowed us to spend any real time together during the week. Having gone to a school at home with less than 200 students did not prepare us for what we were about to experience.

The next few days were a mind-blowing experience. It did not take very long to find out that Maggie and I were among a small group of the youngest student on campus.

There was no time to make friends or do anything other than attend classes, do homework, eat, sleep and do it all over again the next day. Even the classes we shared did not really allow Maggie and I to spend any time together. Evenings were spent at the library or in the dorm doing assignments.

By the time the weekend rolled around, I was tired and a little stressed. Maggie had suggested we take all day Saturday and do something for ourselves without thinking about school. Unknowingly, she had set a pattern that we would use all during our first semester.

If there was not a home football game, we would spend our time exploring the countryside, going to a movie, or sometimes both. Sunday was a day where we called home and did last minute prep for the upcoming week. We very seldom spent time together on Sundays.

The weeks rolled by and we slowly got comfortable with the vigorous schedule of college life. There was very little time for being homesick. But as Christmas break was approaching, both Maggie and I were more than ready to see our families again.

Once a week, usually on Sunday afternoon we received a phone call from Andrea. She was just checking to make sure all our needs were met and that the classes were going well. We had signed a paper which allowed a copy of our grades to be mailed to Rex. He was very aware of how we were doing in school.

Fortunately, both of us were doing well and our grades reflected our dedication. The week before the Christmas break, we had important tests in almost every class. Every spare minute was spent studying. It was the hardest week we had experienced up to this point in our college experience.

Maggie and I decided that we were just too tired to leave after class on Friday. Many of the students who lived closer to school than we did, could make it to their homes in just a few hours. For us, the almost ten-hour trip was not going to start until Saturday morning.

We were up early, and it took only a few minutes to load up the truck. We did not even eat breakfast because we wanted to get down the road toward home. It was so nice for us to be alone and not have any school pressure. We could not have been happier.

Brunch was a very quick stop. We ate quickly as we wanted to get home. I drove for hours without either of us saying anything. Maggie was sitting beside me with her head on my shoulder. For the first time in months, we were free to just be us. Maggie asked at one point if I thought that our families would think that we had changed. I had to think about that question. My answer was that we were still the same and they would see us that way.

When we were less than one hundred miles from home, the level of our excitement was increasing with every mile. It had only been three months since we left home. It seemed much longer.

We arrived at Maggie's home, and she did not even unload anything. She ran for the front door. Her mom must have seen us drive up because she opened the door and wrapped her arms around Maggie. Andy was in town somewhere, so it was just the three of us. I carried one of Maggie's suitcases up to the door and received a long warm hug from Stephanie. We finished unloading everything and I headed to my home.

Stephanie must have called Grandma because, she was sitting on the porch as I drove up. I did not realize how much I missed here until I was holding her in my arms. We just stood there and held each other for a long time without saying a word.

Our moment alone was interrupted, by Grandpa saying, "What about me?" I let go of Grandma and hugged him tightly. He hugged back and patted me on my back.

We unloaded the truck and sat down to a wonderful home-cooked meal that Grandma had ready for us. The were a thousand questions and as many answers as I could give. It was a wonderful reunion.

Grandpa told me that it would be hard for me to imagine how fast Rex was making changes and improvements to our property and the Amos land. He said we could go for a ride and see everything in the morning.

It was too cold to sit on the porch after dark, so we sat around our wood stove for a little longer before I excused myself and went to bed. Before falling asleep, it occurred to me how glad I was that my grandpa had not gotten rid of our wood stove. It made our house a home.

I slept like a baby and woke up to the smell of bacon and biscuits. At breakfast, Grandma said that Stephanie and Maggie were coming over later and that she and Stephanie would give us a tour of the upgrades to the plant.

After breakfast Grandpa and I took a truck and started our tour. Grandpa said we should look at the Amos property first. As we approached the driveway to the ranch, I could see piles of lumber and fence posts everywhere. Grandpa said that the crew was off for the weekend. But every other day they were building fences for the new corrals.

I could see a new building in the distance. Grandpa said it was a barn with stalls in it for the horses that would be used to herd the cattle. We looked over the new pipeline from the largest well. It would distribute water to all the corrals. It was hard to take in the changes that were already taking place on the Amos property.

We finished that part of the tour and headed back to our farm. Grandpa stopped by our main plant and said that we should walk around to the back of the building.

When we rounded the corner, I could not believe what I was seeing. There was an addition being built onto the existing building. The new part would increase the building size by almost three times the original size. Grandpa could see my surprise and he smiled before speaking. "Gule, Rex is not the type of man that goes slow at anything. He is on a mission to make this company a huge part of Walker Ranch."

From where we stood, I was able to see that the Grandma's Pies building was being doubled in size. All I could say was "Wow".

Grandpa said that not much had changed inside either

building because there was ongoing construction. Just after noon, Maggie and her mom showed up with a lunch that Maggie had helped prepare. It was so nice to see her without the pressure of school and schedules to keep. We ate roast beef sandwiches, deviled eggs and coleslaw. It was very good, almost too good. I did not want to make Grandma feel bad, so I did not say too much. Grandpa had no such concerns, so he went on and on about how good it was.

After lunch Stephanie led the way to Grandma's Pies, and we were able to see a few improvements inside the original building. The prep area had been enlarged, and the packaging line had several additions of new equipment which helped speed up production.

Stephanie told us about all the new things that addition would be housing. Grandma said that Rex felt with all the new equipment we would be able to triple our output.

When we were finished with the tour, Grandpa told me that I should take Maggie over to the Amos property and show her around. It did not take me long to agree. We ran to the truck, laughing and enjoying the moment.

When we parked at the Amos house, Maggie grabbed my head and started kissing me all over my face. I guess she had missed me. I returned her affection, and we just sat there in the truck and held each other. It was several minutes before she let me go and said, "Are you going to show me the ranch or not?" I laughed and we started our tour holding hands and just enjoying our time alone.

About an hour into our visit, Maggie said," Gule, do you think we will be able to fit into this huge thing Rex is building?" I told her that we had one advantage. We were born here and grew up with these people. I told her it was going to be an adventure, and we could share it together.

After exploring the Amos farm, Maggie said, "Can we go visit our special place at the lake?" I told her it would be chilly, but if that is what she wanted we would do it.

We spent an hour there before we were so cold that we agreed to go back to the house. The three months at college had changed us from high school kids with undetermined goals to young adults with our future all planned out.

When we got inside our home and were sitting around the wood stove, Grandpa came in and said that he had received a phone call from Rex. He was going to fly up on Thursday morning. He wanted to have a nice lunch for all the employees and give out the bonuses himself.

There was not much time to get ready for the company luncheon. Maggie, Grandma and Stephanie handled all the meal preparations. Andy and Grandpa worked out the schedule of events, and I helped with whoever needed an extra hand.

Rex was expected in the morning. Someone had built a grass landing field during the war, and it was still used by local flyers. Rex had checked it out and was going to use it tomorrow. Grandpa was going to pick him and his crew up when they landed.

The rest of us were busy decorating and doing last minute things to make this a great event for our employees. Maggie and I had not seen Rex for several months and we were looking forward to him attending the celebration.

Everyone was up early, and we had a quick breakfast prepared by Grandma. Stephanie was taking pictures of all our hard work. It really felt like Christmas. About ten o'clock, Grandpa received the expected call from Rex. He and Andy left for the little airstrip.

Within an hour they returned with Rex and four more members of this company. We were waiting for them outside the main door to the plant. Rex saw Maggie and I and headed straight for us. He gave Maggie a huge hug and shook my hand as he patted me on the back.

We went inside and Rex got to see our preparations for the day's activities. He must have been pleased because he just kept smiling and nodding his head. I remembered seeing the other members of his group but could not remember their names. Rex must have read my mind because he introduced everyone as though it was the first time we had met.

Andy asked him how his flight had been, and Rex told him that it was all good until they tried to land. He told us that his company had acquired several airplanes for travel and managing the ranch. He had decided after landing that his crew would build a better landing strip on the back of the Ramos property, if that would be allowed under the lease with Andy and Grandpa.

Andy just nodded his approval. Grandpa however, just rubbed his chin and finally said that he thought it would be ok if Rex would give us all a short flight over our land before he left this afternoon. Rex laughed and said that Grandpa drove a hard bargain. So, it was settled. We would all get a ride on the plane before Rex headed home.

The time was slipping by and Rex asked Grandpa and Grandma to round up all the employees because he wanted to get started a little early. One of his partners was carrying a large suitcase with him. Rex asked Grandpa to set up a table for the suitcase.

It did not take very long to shut down the plant and pile the tables with all the food that had been prepared for our little companies. Grandpa had a couple of the workers help him bring in a small Christmas tree that had already been decorated.

Everyone was smiling and it was like a huge family reunion. Some of these folks had been working for us from the first day we started our business.

When everyone was seated and we were ready to eat. Rex thanked everyone for coming and said he would like to say a few words after our meal. He asked Grandpa to say a prayer before we started eating. Everyone seemed to be enjoying the meal and the anticipation of getting their bonus. It was like one great big family.

When everyone had finished and the dessert was served, Rex stood up and said that he wanted to thank everyone for their contribution to helping the companies have another profitable year. He said that there were many things in the planning stage which would take the companies to a new level and none of them would ever work without the dedicated employees that the company was blessed to have.

Rex had Grandma and Grampa come up and as they read each employee's name, they were asked to come up and get their bonus. I had witnessed this scene for many years during my life, but I never tired of seeing these wonderful folks be recognized for their hard work and loyalty. Without them, there would not be a Meadors & Scott or even the Grandma's Meat Pies.

When each person had their bonus, Rex said that he was sorry to leave so soon, but he had obligations back in Texas. He thanked everyone again and told them that they could take the rest of the day off and the following day as well, with full pay. There was an audible cheer and lots of clapping. Rex asked if we could go to the office that he had a few things to go over before our promised plane ride and his departure.

When we were all seated in the office, Rex thanked the

parents for all their hard work amidst the construction and changes taking place. He said that all his partners were happy with the way things were going in the companies and felt that we were on track to making a big impact on the direction that was planned for the future of Walker Ranch.

He then gave each of them an envelope which we later learned was a bonus. Then he looked at Maggie and me. "I want to thank you two for showing me that my decision of including you in the future of these companies was a good one. Your grades are very good. Andrea keeps me informed about your progress. I am very pleased." When he finished there was an envelope for each of us.

Then he smiled and said, "Now let's go take a plane ride." With that, we were off to the little airstrip where a beautiful two engine plane sat waiting. While we were waiting for the crew to start the engines, Rex told us that air travel was going to change how many things were done on Walker Ranch.

The airplanes and helicopters were going to allow aerial surveys of the cattle, the ponds and lakes, and to check the health of pastures. He was going to be using planes to visit our companies on a more regular basis.

As he finished, we heard the engines start up and we taxied to one end and turned around for our takeoff. For Maggie, Grandma, and me this was our first time in an airplane. As we left the ground, it was the most surreal feeling. Rex had the pilots take the plane over our properties several times. It was amazing to look down on our land, company buildings and our lake. Maggie and I could even see our special little table sitting beside our favorite oak tree.

Stephanie had brought along a camera and was taking pictures constantly during our little plane ride. Rex seemed to enjoy sharing this moment with us because he never stopped smiling.

When we landed, he hugged every one of us. This was new because he had always shaken hands with Andy and Grandma. It was like we were becoming his family. Years later, I would look back on this day with the fondness of memories.

As the plane took off, I think everyone of us was thankful for the man inside that plane. The man who had changed our lives forever and somehow cared personally about each of us. We were very blessed.

The ride back to the farm was very quiet. Each of us was processing our own thoughts on this day. No one opened their envelope. We just sat there and reflected on our new lives. The silence was finally broken by Grandpa who said, "Well there are not many like that man." I think he spoke for what each of us was thinking.

The next few days were very pleasant. Both of our families were preparing for our own Christmas celebration. Grandpa wanted us to spend Christmas Eve and Christmas day alone. He was right in his thinking. We had a wonderful time with our little family. I felt blessed with all Rex was doing for us. When I looked around the room on Christmas day, I could see a home filled with love. I saw the two people who had given me of themselves my entire life.

They were getting older, but not to the point where it slowed them down much. I truly believe that Rex had inspired them with all his plans and given all our lives a new purpose.

On Christmas morning, each of us opened the envelope that Rex had given us. Grandpa's gift was a plan where all the remodeling to the Amos house would be paid for by Walker Ranch. Grandma's gift was her cook stove, and the wood stove would be relocated to the new house, and a new electric washing machine would be installed as well.

My gift was a new dock to be built on our lake and a new fiberglass rowboat. He was also going to build a small patio with a roof, which would house our special table. I could not wait to tell Maggie.

The week between Christmas and New Year's Day went quickly. There was a lot to do at the companies. Maggie and I did whatever was asked of us. Our families did spend New Year's Eve together and that is when we shared what Rex had given each of us.

Andy loved to hunt, but he did not have much opportunity in the years during the war and after. Rex told him that he would be flown up to the ranch and allowed to have a guided hunt in the fall of the year. Stephanie was given a one-week visit to a spa in Texas at the same time Andy was hunting. Maggie was to pick a foal from all the newborn horses in the spring. It would be raised on Walker Ranch and trained so it would be ready for her to bring it back home when it was a little over two years old.

It is hard to understand how this man who was running an empire, could take the time to pick such perfect gifts for the members of our families. In a few short days, we were headed back to school.

# CHAPTER NINETEEN

The second semester was not as eventful because we were used to the routine. Maggie and I still had most Saturdays together. It was business as usual until spring break.

Andrea had informed us that Rex wanted us to visit the ranch during spring break, and he would send a plane to pick us up. It was going to be a wonderful escape after several months of study. I was a little reluctant as Grandma and Grandpa were looking forward to seeing me. We had a long phone conversation that Sunday and they both encouraged me to take advantage of such a generous offer.

On the day the plane was to show up, Maggie and I drove to the airport. The last few months were like some kind of make-believe life. We were almost done with our first year of college. One of the largest cattle ranches in the country had become our employer. This trip on an airplane underlined the changes in our lives.

We reached the airport and after some directions from the staff, we were directed where to park and where to meet our plane.

We sat in a small terminal looking out the glass windows. In less than thirty minutes a different plane than the one we had used to visit our ranch rolled up just outside the building we were in. One of the workers in our area approached us and let us know that this was our plane. We were ushered out by the staff and met at the plane by one of the pilots who gave us a ride at Christmas.

He welcomed us and we boarded the plane. There was a ground crew refueling our plane for the return trip to Walker Ranch. While we were waiting Arnold, our pilot engaged us in some friendly conversation. He showed us around this smaller plane and even let us look in the cockpit.

There was a small locker that contained snacks and drinks. Maggie's parents had given her a camera at Christmas, and she was ready to document our plane ride with pictures. It was surreal. We felt like some kind of royalty. How could all this be happening to us? It was unnecessary to say, we enjoyed every minute of the flight. Maggie took so many pictures that she had to reload the camera three times with fresh film.

When we landed at the ranch, Andrea met us and took us to the cabins where we unpacked our things. Maggie's was right next to mine. Andrea had told us that we had the rest of the day to do as we liked, but in the morning, Rex wanted to see us just after breakfast.

After Andrea left, we went to Maggie's cabin and just crashed on her bed. We held each other and just enjoyed the moment. No school. No list of things to do. Just a wonderful few days to relax with someone who made your world a better place.

The life at school was slowly replaced by the peacefulness of our current situation. I think we both dozed off for a little while. Maggie woke up first and I was awakened as she was trying to quietly move off the bed.

We talked for a while and decided that we would skip lunch and go swimming in the pool, then explore some more of the ranch until supper time and after dinner we would go back to my room and watch TV.

We had a wonderful time and by the time we had dinner we were hungry enough to enjoy many of the items offered for the evening meal.

After dinner we were walking back toward the cabins when Maggie said, "Gule, this is such a beautiful evening, I would rather not spend it inside looking at a TV. We should take a walk and find a nice place to sit and watch the stars." And that is what we did. Looking back at our lives, this was a defining moment where we were getting more mature and our feelings were growing stronger for each other.

It was a wonderful evening, and we stayed outside for a long time. Not knowing what the schedule was for the next day, it was decided that we needed to get some sleep. I walked Maggie to her cabin and after some kisses, I reluctantly let her go inside alone.

The next morning, we walked over to the Chuck Wagon for our breakfast. I had a wonderful night's sleep and Maggie said she did as well. We were enjoying this time at the ranch. The food was great, and we finished and walked over to the main office.

We went straight to Andrea's office, and she was ready for us. " Rex is on his way over and you are going to spend some time with him today" she said.

She had barely finished speaking when Rex came through the door. He hugged us both and asked if we were ready for some fun. I do not think I had ever heard him make a statement like that one. We both answered that we were ready.

He had his pickup outside, and we loaded up for a trip to somewhere unknown. He drove for a while, and we could see one of the larger buildings that had horse stalls and a large tack room. We had visited here briefly on our first tour of the ranch

We entered the building and Rex told Maggie that each stall had a mother and her baby. He told her she needed to pick out the baby she wanted and start spending some time with it.

Maggie was beyond excited. While Maggie was looking in the first stall, I counted twenty stalls on each side of the long barn. She was going to have forty babies to choose from. While she was looking, Rex walked with me through the barn to the opposite end.

He wanted to know if Maggie and I were still committed to the long-term plan. I assured him that we were both on board. That seemed to make him happy.

He looked back down the first row of stalls and Maggie was not even halfway finished on that side. He laughed and said that it was going to take a long time for her to decide.

It was apparent that Rex had something on his mind. He was not doing well with small talk. I finally asked him what was on his mind. I do not believe that I was prepared for his answer.

He gave a long sigh and then said, "Tommy, my wife Abby and I never had any children of our own. I do not know why. I think we would have been good parents. We were talking the other night about life and the future. My partners have great plans for the ranch and other businesses that we own. I am in total support of those plans. But there is just something missing that I cannot put my finger on at this time."

" You and Maggie are not the first young people that we have sponsored in college and offered jobs in our corporation. There are several children of our ranch foreman who have completed the program and now serve the ranch in one way or another. For the most part, it has always worked out for the best."

"My wife and I just feel that something is missing. I want you and Maggie to meet Abby before you head back to school. So tonight, we would like you both to have a private dinner with us at our home."

I told him that it would be our pleasure to have dinner with them. He seemed relieved to have talked with me. He began to smile and be the Rex that I knew. We walked back into the barn.

Maggie had just finished the first twenty stalls. She was more than happy. Rex asked her if she had a favorite so far. She said there was one, but she wanted to see all the others before making her decision. Rex told her to take her time.

We watched her for a few minutes and then Rex said that we would have time to visit the blacksmith shop next door. We walked over and stood outside watching the activity inside the building. There were at least six men shoeing horses and a couple others working on metal projects.

I don't think Rex cared what we were doing, he just wanted to talk. It was very strange to me that this man who managed an empire, wanted to spend time with me. What an interesting man. He had the money and power to change lives, yet he had a warm and thankful heart. It seemed with my limited life experiences, to be a rare combination.

I do not know how long we talked, but Rex suddenly wanted to check on Maggie. When we entered the barn this time, she was almost to the other end of the second row of stalls. We walked to the other end and waited for her. Rex was excited to see her choice and so was I. Maggie was just coming out of stall number 39. Rex asked her how it was going, and she said she had narrowed it down to three choices. We stood and watched her as she entered number forty.

This stall had a beautiful light brown mother with a black mane and tail. Her baby was a little girl or filly. She was a carbon copy of her mother but just smaller in size.

Maggie walked in and crouched down in the bedding. The mother looked her over, and to our surprise, that little baby walked right over to Maggie. Slowly Maggie held out her hand and that baby sort of smelled her and then nibbled her hand. Maggie was able to give her and little scratch on her neck without spooking her. Within a few minutes the deal was sealed. Maggie was in love. It would be the little horse in stall number forty. Rex did not even hesitate, he just said, "Well what are you going to call her?" Maggie thought for a minute before she answered, "Her name will be Precious."

Rex told her that in the morning, one of the trainers would have a sign put on the stall door with the little horse's new name. Maggie sort of choked up and threw her arms around Rex as she said, "Thank you, so much." That totally surprised Rex, but he returned the hug and said, "You are very welcome."

Maggie was walking on a cloud as we left for the truck. Rex told her about having dinner with them that evening. I don't even know if she heard him. She was off in wonderland over her horse.

Rex dropped us off at our cabins and said that he would have someone pick us up at six o'clock. We thanked him and went into Maggie's cabin to decide what to do until the evening.

What a day it had been, and it was not even noon. Maggie's first words when we stepped into the cabin were, "Do you like her? Isn't she the most beautiful thing you have ever seen? Is her name ok? Do you think she will like me? I had to hold up my hand and tell her to slow down and that I did think the name was perfect and she was beautiful. Maggie threw her arms around me and kissed me all over my face as she said, "I love you."

Well, that was new. She had never said that before. I wanted to believe that her emotions were just running high because of her horse. But there was a part of me that hoped it was not just because of the horse.

She let go of me and said she must call her mom and tell her about the horse. I told her that she should do that and when she was finished, I would meet her at the Chuck Wagon for a little snack to tide us over until dinner.

When I walked into the building, there was a group of young people waiting in line for lunch. To my surprise, the group included our student guides Darren and Janie. We had seen very little of them during the second semester.

I did not want to intrude on them and their friends, so I just sat down in the lobby to wait for Maggie. Darren saw me and waved for me to come in and join them. He introduced me to the group. Janie asked if Maggie was here with me. I told her she was making a phone call and would be there shortly.

Darren asked why we did not go home for the break. I was not sure how to answer him. This could be a delicate thing with the new kids getting special treatment. I decided the best way to approach it was that we were catching up on the plans for the expansion of our company back home. He seemed content with my answer.

By the time we had all gone through the line and sat down with our lunch, Maggie showed up and Janie went to greet her. When she had returned to the table with her plate, I decided to try and send her a message "Darren asked me why we were here and I told him we were finding out about the expansion plans for our company." Maggie gave me a questioning look right before I saw the moment she understood. She said, "Yes there has been a lot to discuss and very little time to play."

She took the empty seat next to me and I grabbed her hand under the table and gave it a loving squeeze. The conversation moved away from us and the members of the group talked about things they were known to all of them.

That made it much easier for Maggie and me. I was hoping she would not say that we were having dinner with Rex and Abby.

The rest of lunch went well with a few questions about our hometown and our company. They were a nice bunch of kids, and I did not want to say or do anything that would make anyone feel we were getting special treatment. I still did not even know myself. But it was becoming obvious that there was something more that was going to be revealed.

As the group broke up, we thanked them for including us in their lunch and then we left for the cabins. When we got settled into the airconditioned room, I flopped down in a chair and told Maggie I was tired. She was still flying high over her horse. I told her that we needed to be very careful about things between us and Rex. I did not want to create jealousy or hurt feelings. She totally agreed.

We talked for some time about why we were getting the special treatment. Things like a private dinner with Rex and Abby were not normal things for the other kids on the ranch. Getting to pick a free horse was probably not an everyday happening. So why us? Maggie said she did not know and did not care. She was loving every minute and wished we never had to leave this wonderful place.

We decided that we had time to go for a swim and then get a short rest or even a nap before our ride came to take us to dinner. Maggie did have it right. This was a magical place. The workers all seemed happy, and the ranch was perfect.

I kept some of the talk with Rex to myself. I really did not even know how to explain it to Maggie, so left unsaid seemed the right thing to do. We went swimming and went to our own rooms so we could rest and get ready for dinner.

## CHAPTER TWENTY

Our ride showed up at six and we were driven to a part of the ranch that we had not seen before this evening. It was about three miles from the headquarters and sat on a little hill overlooking the area. As we left the car that delivered us, I just stood and looked at this paradise below us. Cattle and pastures as far as the eye could see. Peaceful and perfect would be how I would describe it.

As we were taking in all that we could see, the front door opened and both Rex and Abby came out to greet us. Rex had a huge smile on his face. I think he was able to see that we were totally take by the view.

Both he and Abby greeted us with a hug. I had no idea what the evening would include but I was already stunned by what I had seen so far.

We were escorted inside, and Abby gave us a tour of the home. It is not an exaggeration to say that I had never been in a house that was so elegant yet felt so much like a home.

There were huge glass windows so the view could be enjoyed from almost every room. The kitchen would get the blessing from the best of gourmet chiefs. The great room had deer heads mounted on all the walls. A giant fireplace was made of rock and was the center pierce of the entire room. There were shelves along one wall that held Native baskets and pottery.

Throughout the home were paintings of western scenes. Cowboys, cattle, horses, Native Americans, and old cabins, adorned the walls. It was breath taking.

It must have been amusing for Rex and Abby to watch us try and contain our wonder of this place. When the tour ended, Abby led us to the dining room where we sat down for our dinner. This room was as elegant as all the others.

The table could easily seat twenty people. Rex told us it was made from some trees that grew along a river that ran through the ranch. There was a huge cabinet that looked as though it was made from the same wood. It was full of beautiful plates, platters, and crystal.

The magic of the moment was broken by Rex saying, "Let's have a prayer before we start dinner." As soon as he finished, two ladies started bringing out our dinner. I do not think that I can even remember what we ate. It was such a perfect setting, and we were sharing the evening with two of the most special people in our lives.

Both Rex and Abby shared some of their own backgrounds with us. It made the evening more personal, and it was one that Maggie and I would talk about for many years.

Abby asked a lot of questions that were just designed to get to know us better. Rex allowed her to lead the conversation as he had already spent time with both of us. They were people that talked with you in such a way that you felt trust and that you could tell them things that you rarely told others.

As Abby and Maggie talked, I thought to myself, how sad it was that some boy or girl was missing out on having these two for parents. Then I realized that they were missing out on having a child of their own. It was beginning to dawn on me what Rex was trying to tell me while we were waiting for Maggie.

We had a wonderful evening. As we were leaving Rex said, "You only have a couple of days left, so Abby and I would like to have another dinner with you the night before you leave". We assured them that we would look forward to being with them again.

Rex made a call and our ride back to headquarters

showed up in a few minutes. As we were heading for the car, Abby stopped us and gave each of us a big hug. Rex did the same. Why were these people treating us a family? It was a question that Maggie and would spend many hours discussing.

When we were dropped off at the cabin, Maggie said, "Gule, could we go down to the barn so I can check on Precious before we go to bed?" That is what we did. I got a truck from the parking lot, and we drove down.

The barn had lights down along the center walkway. The stalls were in the shadows of those lights. Number forty was just inside on the right. Maggie peaked over the gate and inside the stall both mom and baby were still awake. She carefully opened the gate and moved inside. It was the craziest thing. That little horse walked right over to Maggie.

I watched as the two of them touched each other. It was a very tender moment. There was no doubt this was a love affair that would last for many years. Mommy just watched Maggie petting her baby. She seemed to trust Maggie after such a brief time. I could not believe it. We had horses on our farm, and they did not typically trust strangers.

I could not remember ever seeing Maggie so happy. As we got ready to leave the little filly allowed Maggie to hug her. She even followed Maggie to the gate. I was standing outside watching as Maggie said goodnight to her baby. As I moved out of the way so we could close the gate, I noticed a new sign on the stall. I simply said "Precious".

When Maggie had closed the gate, I pointed toward the little sign. Her eyes teared up and she just hugged me and would not let go. After wiping her eyes, she looked at me and said, "How could I already love her so much?" That question did not require an answer so just squeezing her hand seemed appropriate.

We drove back to the headquarters without saying a word. It had been quite a day. It seemed that Maggie and I were drained of emotion. This ranch and the folks who owned it were much more than our employers.

We walked back to the cabins hand in hand. Maggie said, "Gule, I hope you understand. I am just drained. I want to sleep for two days." I told her it was fine that I was tired as well. We had a warm kiss and embrace as we parted for the night.

I slept like a rock and could not believe how late it was when I woke up. After taking a shower and dressing, I knocked on Maggie's door. She opened the door and was totally ready. "Can we go see Precious?" I asked her if we could go get breakfast first. She laughed and said that would be fine.

We had the day to ourselves. We made a trip to see her filly, went swimming and went back to see Precious and then had dinner. Darren and Janie stopped by our cabins and offered to take us on a trail ride with some horses the next morning. We accepted and set a time to meet for breakfast.

We sat on the porch to watch the stars and talk about our last evening meal the next night with Rex and Abby. Maggie said, "I wish we did not have to go. I am going to miss Precious." I felt the same way. Life away from this ranch did not seem very inviting, at this moment.

Maggie fell asleep on my shoulder. I watched her as she slept. Her red hair framed that lovely, freckled face. I wanted her to look at me the way she did to her little filly. After perhaps a half hour she stirred. I told her she needed to get up and go to bed because we had a big day tomorrow.

She did not protest, kissed me goodnight and went inside. After going inside my cabin, it did not take me long to get in bed and fall asleep.

The next morning, I stopped by her cabin and Maggie, and I walked over to the Chuck Wagon to meet Darren and Janie. They were already seated and eating so we hurried through the food line and sat down with them. Darren said, "So we heard this is your last day on the ranch." We did not know how that they knew, but we just said yes and told them we would miss everyone.

Janie spoke up and told us a little about the trail ride. She recommended that we get some extra food for a lunch stop. "The kitchen staff will pack it for the trip." Everyone picked out lunch some items and we left for the stables.

Darren had made all the arrangements, and the horses were ready when we arrived. We loaded our drinks and supplies in the saddle bags and headed out for what we were told was a smaller lake a few miles east of the headquarters. We let the horses pick their own pace and an hour later a small grove of oak trees indicated the lake location in the distance.

It was so nice. There was a perfect picnic area in the shade of the trees. There were cattle grazing all around the lake. I was so peaceful. Darren said we could let the horses go and they would not wander off. We spent almost three hours eating, talking and wading in the lake.

Janie and Darren had a lot of questions about our future plans . We were careful to not be very specific beyond just working in our family companies that Rex had purchased. Our answers seemed to satisfy their curiosity.

I finally had to suggest that we head back because we had a few things to take care of before leaving in the morning. Everyone was ok with that. We had enjoyed our time together.

We arrived at the stables and unloaded our supplies. One

of the workers took the horses and we were free to go. Maggie thanked everyone for the wonderful time, and we left for the cabins.

It was mid-afternoon by the time we settled into our cabins. Maggie said she would like to take a little rest and a shower before we had to leave for the Walker's home. I agreed and told her I would be ready by the time our ride showed up.

At exactly six, our driver parked in front of the cabins. This being our second trip to the main house, had his interest. He was more talkative. We learned his name was Warren and he was the main driver for Rex and Abby. Neither of us revealed relationship with his boss. We just used our now usual line that we were going to work for him in our family business that he had purchased.

When we arrived for our second visit, no one greeted us outside. We walked up to the door and knocked. Abby opened the door, and we had what was becoming normal, the hugs. Rex was waiting for us in the great room where we received more hugs.

The evening was just as interesting and special as the first time we visited. But that would all change after dinner. When we were all seated in the great room. Rex said, "I wanted you to spend these times with both Abby and me so she could see what I have been telling her since I first met the two of you."

"We have spent a lot of time sharing our thoughts and feelings concerning your futures. I know that we talked about you graduating and then Maggie running Grandma's Meat Pies and you running Meadors & Scott. I still feel that you would do very well in those positions. With your leadership both companies will have a bright future."

He hesitated before he spoke again. After taking a long

breath he started to speak again. "Abby and I have been thinking of the future of the ranch. It is true that it is the heart of all our business interests. But it is more than that. It is a special place where our hearts are invested. We have a commitment to this land, our animals, our workers, and that carries a burden of responsibility."

"We have been searching for a long time for some next generation leadership for our companies. Recently we came to realize that although we were finding individuals for positions that will support the direction the ranch is taking for the future, we have neglected the leadership of the ranch itself."

"Abby and I are the caretakers of something much bigger that an individual or a family. This ranch is home to many families. They are the ones who make this ranch possible. But there must always be leadership. Leadership that encourages others to grow and prosper. Leadership that never forgets that this ranch is a joint effort. Leadership that encourages the talents of those we employ."

"Most importantly the leadership must have a burning in their soul to protect and to be loyal to the land. We have several hundred employees that enjoy working here. They love the land and our way of life. Those qualities alone, do not make them leadership material."

"We want to offer you an opportunity for the coming summer. Abby and I want you to spend the summer working on the ranch, not as leaders. You would be workers who are learning from others who already know their jobs."

"When the current semester ends, you could drive home for a week and enjoy a visit with your families. I will send a plane to pick you up so you can spend the summer here. A week before the fall semester starts, we will fly you home for another visit."

"It would be a chance for you and Maggie to see if there is that burning in your soul for this ranch. I am sorry to bring this up so late in your visit, but I wanted Abby to see you and make her own decision. We are together on our opinion that we would like to see if there is a future here for both of you."

Neither Maggie nor I knew what to say. This was so unexpected. It was unreal that this was happening. Maggie asked Rex if it would be ok to ask some questions. He smiled and said, "Of course."

Maggie told him that it had been somewhat of a burden to us when we were asked why we were here or why did we have dinner with you. She said that we did not want others to feel that we thought we were special or above them or that we demanded special treatment.

Before answering, Rex looked at Abby and smiled. "Maggie these are the qualities that show you are special. You are thinking of other people's feelings and their perception of what they do not understand. So, to put your fears to rest, I will tell the necessary people that Maggie will be working at the Chuck Wagon and that Tommy will be working as a normal ranch hand."

"This will get people comfortable with you both, as you will be employed as one of them. Beyond that, any personal plans or conversations with Abby and me should stay between us. All they need to know is that you will be the new leadership of your family companies when you graduate."

"That is the truth because we have no firm plans beyond those. Go back to school. Concentrate on finishing the semester. Let me know your decision, one month before the school year ends. That way there will be time to make plans. We can make one upgrade for you by allowing each of you to have a cabin for the summer."

I had not said a word up to this point. Abby asked me if there was anything that I would like to say. I thought for a minute before I spoke. "You have given our families a once in a lifetime opportunity to leave a legacy with our little companies. You have allowed us to have a place for our employees to work as long as they want. And you have made our parents feel worthwhile. That alone is the most meaningful thing in my life."

"Secondly, the confidence you have placed in Maggie and me by sending us to school and offering us some equity in your companies is more than generous. There is no way we would ever be able to pay you back for those things. And now you have reached inside our hearts and asked us if we would like to become caretakers along with you of this magical land and the people who have helped you build and maintain it."

"I am not sure that a seventeen-year-old boy and girl, have the maturity to see themselves the way that you do. It is too much to process. I can only say that it is very humbling and an honor to be any part of your plans, for the future. Maggie and I will discuss this more times than you make think. We will make our choice by your timeline."

"You may not know it, but I am an orphan. Grandma and Grandpa were just a kind couple who took me in when I was a baby. They are not blood relatives. They are the only family that I know. In this short time that we have known you, I feel those family feelings about you. Thank you for helping make these changes in our little world."

I finished and did not want to see any reaction to my words , so I just looked down at my lap. Rex waited for a moment before he spoke, "It has been our pleasure to have you spend time with us. I do not want to hold you up."

"Maggie, Precious will get the best treatment of any horse on the ranch. Have a good trip back. Let us know if you need anything." There were some very long and meaningful hugs as we were leaving. Abby even gave me a little kiss on the cheek. Rex saw that and gave me a wink and a thumbs up.

When we got back to the cabin, Maggie wanted to go see Precious. We spent almost an hour there and as we left, I saw some tears running down her cheeks. This place had a magical effect on both of us.

We spent the rest of our last night on the porch of Maggie's cabin. She told me that when I was telling Rex about being an orphan and how they felt like family, that Abby was wiping her eyes.

Years later Maggie and I would look back on these few days with the same wonderful feelings as we were sharing that evening. We sat for almost an hour just holding each other. There did not seem to be any need for words. The moment itself was enough.

Rex had told us that the plane would be ready to leave by ten, that way we would have time to eat and say any goodbyes. Maggie made me promise that we would visit Precious first thing in the morning.

We kissed and held each other just sharing the moment. I went back to my cabin. There did not seem to be any need for words. The moment itself was enough. My emotions were running too fast to allow me to think about issues.

In the morning, we ate a quick breakfast and then stopped by Andrea's office to say goodbye. She was glad we took the time to see her.

The next stop was the stable. Maggie had brought along her camera and wanted me to take pictures of her and Precious. When she opened the stall door that little horse ran right to her. It did not make sense to me how they had bonded in just a few short visits. But there was no doubt that they were special to each other.

We did not have a lot of time before we were to leave so Maggie brushed Precious and talked to her as I took pictures. She hugged that little horse and kissed her goodbye, and we left to meet our ride to the plane. Maggie did not cry, but I could see she was trying to be brave. I think in many ways we were both making that effort.

As the plane took off, there was a strange feeling in my heart that Maggie and I were leaving something important behind. The ride back to school was not eventful. We did not take pictures. We did not talk about what we had left behind. We sat and looked out the window and were both lost in our own thoughts.

## CHAPTER TWENTY-ONE

As the semester started, it was back to school mode. There was not time to dream. We had classes, tests, and the end of the semester exams were looming. The one thing Maggie had made a priority was getting all the pictures, from our trip, developed.

We had both called home and talked to our family about the days at the ranch. There was no amount of talking on the phone that could adequately describe all that we had experienced. However, we made every effort to not leave any of the facts out as we relived the trip with our parents.

The time was moving quickly. Maggie and I were having a hard time using our Saturdays for alone time. We often would meet after our family phone time on Sunday.

As the deadline approached to respond to Rex, we discussed that we should inform our family about his request and allow them to feel that they were part of the decision. We picked the day we would let them know and each of us called our own family and talked about Rex, the ranch, and his offer.

After our individual calls, we met at the school snack bar and filled each other in on how the call had gone. Our families reacted the same way. They would miss us if we were to be gone almost all summer, but they would support our own decision concerning the offer.

That left it up to Maggie and me to decide. It was so far from what we had planned. When Rex wanted to buy the companies, we were told we would spend the summers working at home and applying the things we were learning at school to our jobs within the company.

Now he wanted us to be new hires working at ground level on his ranch. Neither of us had a problem at working as newbies. We were just unable to see the big picture that Rex and Abby were trying to build.

The deadline was days away and we decided to take a Saturday off from school to talk about the choice we needed to make. We drove out to a park near the school and sat down at a picnic table. We took turns doing pros and cons on all the issues we faced.

In the end the positives outweighed the negatives. We both loved the ranch. We would see each other at some time each day. Maggie could help with training Precious. We would both get better acquainted with the ranch and the employees. The hard parts were being away from our families and the jobs we had each summer since we were little children.

By the end of the day, we had made our choice. We would be going to the Walker Ranch for the summer and maybe for the rest of our lives.

We agreed that we would go back to the campus, eat dinner and then call Rex. Maggie wanted me to give him the news. When he answered the phone, we talked small talk for a couple of minutes. I worked up my nerve and told him that we had decided to work at the ranch if he still wanted us. He said that he was not going to say anything out loud because Abby could hear him. He wanted to put her on the phone and have Maggie tell her. Rex got on a second phone in their home and all four of us talked for a long time. Abby told Maggie that she personally checked on Precious several times a week.

Rex started making plans about what we were going to be doing, and Abby had to slow him down. She said, "Rex, give the kids a chance to think. They have made a huge decision, and you are going a hundred miles an hour." He laughed and said how happy he was that we were going to be at the ranch this summer.

When we got off the phone, Maggie and I felt even better about our choice for the summer. We decided to go call our families with the news. As expected, they all supported our decision and were happy for us but sad that they would not share the summer with us.

With the tough choices already made, we had to knuckle down and finish the semester. Maggie gave me some of the pictures after we had put several of them aside to place in a picture album. We each mailed some home for our families to see.

We made a time to have lunch with Darren and Janie so we could tell them we would be working at the ranch for the summer. When we told them what we would be doing, they thought it would be fun for all of us that summer.

The days went by so quickly. Rex had called and told us that when the plane picked us up, it would be carrying one of the ranch hands. He would drive our truck back to the ranch for us.

It was difficult to understand how that man could organize everything down to the smallest detail. He had already planned our departure day to home and the one from home to the ranch. Rex told us not to work when we got home. He wanted us to enjoy the visit.

Maggie and I finished our semester exams and had a partial afternoon to ourselves. It had been and tiring week. We went out for dinner to celebrate making it through our freshman year of college and then we packed for the trip home the next morning.

The plane arrived at ten and we were there to meet it. One of the horse trainers had been picked to drive our truck back to the ranch. We greeted him and gave him the keys. He said he was going to get lunch and head back. Before he walked away, he turned to Maggie and said, "Precious is doing fine." Maggie thanked him and we got on the plane.

It was so relaxing to hear the engine noise and nothing else. I was asleep within thirty minutes of us taking off. What seemed like only a few minutes later, Maggie shook me and said we were about to land. I could not believe that I had slept that long.

The plane landed on the new gravel strip Rex had built on the Ramos property. We taxied up to one end and Andy, Grandpa, Grandma and Stephanie were standing there waiting. There were lots of hugging and some kissing before we even got into our transportation.

It was well into the afternoon, so we all agreed to meet for dinner the next evening. Maggie and her family headed for town, and I thought we were heading for home. Grandpa stopped in front of the Amos house and he and Grandma got out. I was a little confused. Grandpa smiled and said, "This is our home now Gule."

I had only been in the Ramos house one time before and remembered very little about the floor plan. Grandma wanted to give me the tour. Her personality was evident in the furniture and colors of the walls. True to his word, Rex had our wood stove installed in the room that we were now supposed to call the family room. Her kitchen had new appliances. The bathrooms, that's right bathrooms, had a tub and shower in each one.

Her new laundry room had a sink and an electric washer. The bedrooms were much larger than we had in our old house. There were three of them. Each one had a closet and chest of drawers. Mine had my baseball glove and a ball sitting on top.

When the tour was over, I hugged Grandma and told her I was so happy that she had this wonderful place. Grandpa told me that he did not have anything to show me because he no longer had to feed, chickens, pigs and cows. He was like a city person. He went to work during the week and had weekends off. He said we do not even have a garden.

As he spoke, I looked closely at him. This man who was my grandpa, my father, and my best friend, looked happier and younger than I had seen him in years. Both their lives were so much better than the years of hard work they had been so used to their entire existence.

Rex had been a major factor in making all this happen. But if Andy, Grandpa and their wives had not started our companies, there would have never been an opportunity to meet Rex.

I unpacked the few things that were in my suitcase and went out to visit with the two people who loved me more than anyone else on earth.

We had a wonderful evening catching up on what was happening in our lives. Grandpa and Grandma were going to show me the improvements to their companies after breakfast in the morning.

I know it had only been a few months since we saw each other. To me, it seemed like years. Each of us had our lives altered by the circumstance of knowing Rex Walker.

Even though I had slept on the plane ride home, I was ready for bed when they said goodnight. The bed that I climbed into was also new and it was more than comfortable. It did not take me long to fall asleep.

After breakfast, we drove to our ranch and the additions to the buildings were amazing. We toured Grandma's company first. The entire processing lines had been modernized and enlarged. The was and entire wall of the new addition that was nothing but giant freezers. A very large loading dock could service the loading of five large semi-trucks at one time.

It was very impressive. Grandma knew how every single machine in that plant operated. On the main office wall were two doors leading to offices. A sign on the wall read Plant supervisor Ida Meadors and the other read Plant supervisor Stephanie Scott.

When we got over to Grandpa's side of the ranch, there were many new changes as well. There were several air-conditioned rooms where hanging meat was aged. Freezer that kept meat frozen before shipping lined one wall. The meat cutting department had grown from ten stations to forty. The entire plant had a trolley system that moved the meat from place to place.

Everything was stainless steel to help with easy cleaning. It was like some futuristic place. Grandpa understood how each part of the system worked and how to maintain them.

The last stop was our old home. Rex had completely changed the floor plan and up graded it to an office building. There were several small offices and a couple of larger ones. Each had a phone, typewriter, adding machine, and filing cabinets.

Each office had a name on a brass plate that was on the wall by the door. One said General manager Rex Walker. Another said Plant supervisor Andy Scott. And one more said Plant supervisor Roy Meadors.

It looked very professional. Grandpa showed me the inside of his office. It had a few pictures of our farm before it was covered with buildings and one of me with Ogi. The largest was a picture of him with Andy and Rex. He told me they were having enlargements made some of the pictures that Maggie had taken.

Grandpa said, "We have one last place to see." We got into his truck and drove down to the lake. Before we even got there, I could see a building beside the oak tree. We walked over to it and just as Rex had promised, he had the crew build a beautiful three-sided shelter from the sun in summer and storms in winter. Inside was our special table and the bench that Maggie and I had taken down there years ago. Rex had also had a brass sign made for Ogi's grave.

All the changes were too much to seem real and my emotions were getting to me. When we got to Ogi's grave and that brass marker, I could no longer hold back some tears. I just threw my arms around Grandpa and held him. Grandma came over and we had a group hug that lasted a long time.

The three of us just stood there and enjoyed the simple way our lives used to be. Grandpa said, "I think you missed one more thing. He walked us over to the water's edge. There was a red and white boat tied to a little dock that had been installed in the lake. On the back, painted in big letters was the word GULE. Our farm and the Ramos property were being transformed into another magical place. I guess that was what Rex did turn places into magical kingdoms.

We went back home, and Grandpa and I helped Grandma get ready for the evening meal with the Scotts. The change in the family finances over the years, did not take away from Grandma wanting to cook. It was her joy to prepare meals for those she loved.

When she finally kicked us out of her kitchen, Grandpa and I took a walk out to the area where Rex was having his men build all the new corrals and holding pens. It was a sight to see. Thousands of fence posts stacked up and ready for installation. It was beginning to look like another Walker Ranch right in our back yard.

Grandpa stopped walking at one point and asked me if I had any problem with the direction Rex was taking me. I told him that my only reservation was that I was not going to be working with him in our company. I told him how much that I missed them every single day when I was not at home.

He gave me a hug and said that it was the hardest thing they lived with each day. But they knew they had to let me go and make my own way. He said that God had blessed them with good health, but that did not stop them from growing older. "It had always been our wish that you would find a wife and settle down on our farm, and we could grow old watching you have children of your own."

"I wanted to take them to the lake and help them catch a fish. I wanted to teach your boys how to play baseball. You have made me proud, and in some ways, I am jealous of Rex having time with you." He stopped and had to wipe his eyes.

This was the man of few words. He was expressing his deepest thoughts to me. I was glad he did not ask me to respond, because I do not think any words could have made it out of my mouth. I was fighting back my own tears. Rex cared for me because he saw a place for me, in the future. Grandpa loved me when I had nothing to offer. I was just a baby, and he loved me and raised me. There was a huge difference. I wanted to tell him, but the words would not come.

As we walked back to the house, I was finally able to get out a few words. "Grandpa there will never be anyone who can take your place in my heart." That seemed to make him happy. When we checked on Grandma she told us, "Don't be hanging around and getting in my way. Go over to the plant and see if the Scotts are there."

We both knew better than to protest so we drove over to the plant. The Scott's truck was there so we knew Maggie was getting the tour.

While we were waiting for them to finish, I asked Grandpa who was going to manage and feed all the cattle that would occupy the new pens. He said that Rex was going to build some housing for some of the workers who would be moving up from the Walker Ranch. They would have their own foreman and would not be under the supervision of him or Andy.

Rex had not started construction of the hay barns and stables for their horses. Grandpa said that he wondered if there would be any farm ground left after everything was built and the pastures were established.

Everyone took for granted that with the establishment of our own little airstrip, Rex would be making many more trips to our area. The crew that he had constructing the new pens were staying in town. Rex rented every room in one of the few motels in town.

The town was glad to see all the changes. His projects were bringing a lot of money to our little community. Andy was selling more lumber than he could mill. He had resorted to buying outside lumber to help Rex. The mammoth project that Rex had envisioned was becoming a reality.

Our families got together several times before our week was up. Maggie and I even had time to visit the lake. It was difficult to fully enjoy the time because the knowledge that we were leaving soon was always in the back of our minds. It would be almost three months before we would return. The plans that Rex was creating, chipped away at the time I had to be with my grandpa and grandma.

They were not going to live forever. These were supposed to be their golden years, and I wanted to spend that time with them. The night before the plane was to return for us, I had a long talk with them about my thoughts.

Grandma did not say much. There were many times during our conversation, when she needed to wipe her eyes. Grandpa slowly formed his response to my concerns. "Gule, the Creator has made a pathway for us to travel. When you are young, your family is everything. They love you, teach you, protect you, and wish nothing but happiness for you."

"The Creator loans you to your family for a season. Then He puts a desire in your heart for a mate and creating your own road. It is the way of things. Grandma and I have always known that we would have you for a season. We have been blessed more than you know, to have you as our grandson."

" Now, our wish is to live long enough to see you travel on your new road. The oak tree produces the acorn and the acorn in turn produces a new oak tree. It is the Creator's perfect plan."

He smiled and it was clear he was done speaking. I had no idea how to respond, so I just nodded my head. I was fighting back my own tears and would not have been able to speak even if I wanted to.

Before going to bed, we had one of our favorite things to do. We held each other in a group hug. As I was falling asleep, I kept thinking what Grandpa said about the Creator loaning me to them. I would never forget that statement. It would impact my thinking for the rest of my life. In the morning, we had a feast prepared by Grandma. I think each of us was trying to control our emotions. We did not talk much about things of the heart.

The plane showed up on time. The Scotts were at the airstrip when we arrived. We made a short process of goodbyes. Lots of hugging and kisses, before we climbed into the plane. As we taxied off, the four people who were our world were standing there and waving to us.

## CHAPTER TWENTY-TWO

Maggie and I were quiet for a long time. Each of us was lost in our own thoughts and feelings. I am not sure that it would have been possible for me to articulate my exact feelings. There was some sense of this ranch experience being an adventure. But that feeling was tempered with a feeling that I was missing out on this special time in the lives of Grandma and Grandpa.

I wanted to be able to share the success and financial freedom that they were finally experiencing in their lives. They were so deserving of all they had earned by being hard working, honest and caring for others. Their generosity was well known in our part of the state. Without telling Maggie, I had made a promise to myself that nothing was going to prevent me from spending time with them in the future.

Maggie broke the silence, "I want to go see Precious as soon as we get to the ranch." I smiled at her and assured her that we would head go the stalls first thing after settling into our cabins.

The plane landed and we were met by Andrea. She helped us load our things into a truck and we drove to the cabins. We were given our keys and each of us was handed a folder. Andrea told us we had the evening free, but in the morning, we were to report to the people named in our folder. She told us how happy she was that we had accepted the offer to work on the ranch for the summer.

We unpacked and met Maggie outside her cabin. As hard as it was, Maggie agreed that we should eat before going to see Precious. We had a wonderful meal before hurrying over to the barn.

Maggie said, "Do you think she will remember me? I hope she does." Any words that I might say were worthless. Her heart was on that little filly and nothing else was important.

We walked into the barn, and Maggie tentatively opened the stall door. Both horses looked her way and just stood there. To Maggie's joy, Precious walked over to her and nuzzled her hand. The little horse had grown in the week we were gone.

Not wanting to break the spell, I sat in the corner of the stall and watched Maggie love on her little horse. The time went quickly and before either of us realized it, we had been there for almost one hour.

We walked back to the cabins. Any sadness Maggie had when we left home had been replaced the moment Precious walked over and greeted her. We had dinner that evening and agreed to go to bed early so we would be fresh for our new jobs in the morning. Before we left each other, we opened our folders and compared what we found inside.

Maggie was to report to the Chuck Wagon at 5am. I was to report to the stable where Precious and her mom were housed. My meeting was to be with one of the foremen named Everett, at 6am.

I do not think either of us slept very well that night. The anticipation of the next day's activity was the main reason. Missing Grandma and Grandpa was weighing heavily on me. In the morning, I had breakfast at the Chuck Wagon. Maggie was nowhere to be seen. She was no doubt busy in the kitchen. I ate and hurried over to the stable to meet Everett.

There was a group of men standing by the main door. They all watched me as I walked up. A tall cowboy looking man in his forties was the first to speak. "You must be Tommy." I nodded my head. He stuck out his hand and said, "I am Everett, and these fellows are your coworkers." He introduced each one and we shook hands all around.

He told me that I was to work with his crew for two weeks

we would be moving horses around to different locations on the ranch and helping the horse trainers. He pointed to one of the guys and said, "Mark will take you over to pick out a horse and get it all tacked up and ready to ride."

That was the beginning of my ranch adventure. The next two weeks were a huge learning curve. My knowledge about horses would greatly increase during my time with Everett and his crew. My horse was named Willie, and he loved to work.

I know that sounds silly to someone who does not understand horses. But it is true. Willie knew his job much better than I did. Each morning, he was ready to be saddled, and he could not wait to get out and go to work.

Those two weeks were filled with moving horses between pastures. Taking mothers and their babies from the stalls and putting them in pastures where they could run and eat grass. Putting mares who were about to give birth into the stalls. Taking horses to the blacksmith to get new shoes. Working with trainers to accustom colts to wearing a halter and being led with a rope.

We helped the vet check out the health of the ranch herd. There was some down time where we practiced roping. I really needed this training. After the two weeks, I was still a terrible roper, but I was learning. Willie had figured out that he knew much more about horses than I did. It was almost a game between us. He would wait for me to let him know what he was supposed to do and when I failed to tell him, he just went ahead with the job at hand. He was the boss, and I was his student.

Maggie was doing better than I was. She knew her way around a kitchen and showed her skills to her boss Wendy. Unlike me, Maggie would not be moving on after two weeks. Her summer was going to be working every day in the Chuck

Wagon.

The only time we had to be together was on our day off and they were often on different days. We would sometimes meet in the evening for a short visit. With the morning coming so early, we were too tired to stay up late.

Maggie told me that creating meals with a large variety of food was much different than working at Grandma's Pies, where there were only a few items created repeatedly. Her first assignment had been baking bread, rolls, biscuits, cakes and cookies.

When I was able to see her, she told me she was happy and learning a lot about feeding the ranch staff. She would tell me how she still made time to visit Precious even though she and her mother had been moved to a nearby pasture. The trainers had shown her how to place the halter on her head without spooking her. Maggie spent time leading Precious around the pasture with a rope. She brushed her daily and their bond grew with each day.

The first two weeks went by quickly. Maggie was moved to the meat department and prepared meat dishes for all three meals.

I was moved to a different crew and worked herding cattle from pasture to pasture. We took field counts and did record keeping about new calves born. We placed ear tags on the new babies and did some branding.

I was able to bring Willie with me. As usual, he knew the job better than I did. I was still needing to improve my roping skills and Willie would position his body, to try and help me rope a cow or calf. I gained more respect for that horse every day that we worked together.

Over the years since that summer, I have thought about him many times. I truly believe he was doing his best to teach me how to do my job.

## CHAPTER TWENTY-THREE

The fourth of July was coming up in a few days. Maggie and I learned that Rex put on a great celebration where everyone on the ranch was invited to a huge barbeque followed by an afternoon of fun and games.

The Chuck Wagon staff spent several days creating a feast to go with all the barbequed meat dishes. The rest of us got horses and cattle ready for races, roping events, and games for the children.

The chores on a ranch do not stop because of a day on the calendar. As each group finished their normal chores on July 4th, they would gather at the picnic site. We had readied some of the corrals for games and events. Everyone was excited and it was clear that Rex used this day as a chance thank everyone on the ranch for their hard work.

Before we ate, Rex gave a little speech and then prayed for God to bless our meal. He and Abby helped the staff serve the meals. Maggie was busy making sure the food line was kept full of wonderful choices. I went through by myself.

When I reached the place where Rex was helping serve, he gave me a big smile and asked how things were going. I gave him a thumbs up and said that I was learning a lot about the ranch. He nodded and said, "Keep up the good work, we will be talking soon."

I had a chance to wave at Maggie as I finished the line. It was a delicious meal. It would have been even better if Maggie could have shared it with me. When I walked up to get dessert, Maggie was restocking the cake plate. She saw me and said that she would only have to work another hour. After that she would be free to spend the rest of the day with me.

That was the best news all day. I was not going to be in any of the events because my skill level was far below those of almost every hand on the ranch. It was fine with me because Maggie and I could just be together and enjoy the rest of the day.

Our second two-week period was ended and neither of us knew what we would be learning next. Before the day was over Maggie would learn that she was going to be learning how to plan a week of meals in advance. I was contacted by Brad, who was the water resource director for the entire ranch. He informed me that we would be spending the next month together. I knew who he was but had never talked to him before today.

The rest of the day and on into the evening Maggie and I shared the experiences we were having as we did our duties. What Maggie really wanted to talk about was Precious. She shared that her little horse was now running to her when she approached the pasture where he and his mother spent their days.

When she talked about Precious, there was a way her eyes sparkled. I believe that Maggie would never recover if anything bad ever happened to her horse. My stories about Willie were mildly of interest to her. Maggie lived to do her work so she could finish and then go visit her horse.

The next month was very busy. Maggie was learning how much effort was made to feed an entire ranch. She learned how to order supplies, plan the meals so that they were tasty and satisfying, and were served on time.

My education was far from what I had expected. Water management was far more complicated than I would have guessed. Brad was an excellent teacher.

My first day was spent with him flying over most of the ranch in a plane. He had given me a map of every pond and lake on the ranch and as we flew over them, he would point each one out on the map. Attached to the map was an information sheet which told how many acre feet each impoundment held.

I would learn that all the information gathered would allow the ranch to determine how many acres of pasture could be irrigated that next spring. That in turn could give a picture of how many cows a particular area could support. The ranch had the ability to pump water from one area to another using above ground pipes. But distance was a limiting factor.

The information gathered by measuring the water supply was one of the most important factors in having a productive and financially feasible ranch. The entire month was spent in the field measuring water levels and then returning to the office to record the figures and make projections.

On the days when Brad and I were in the office, Rex would occasionally stop by to see our figures and ask Brad questions. It was always good to see him, and he was always encouraging me with his award-winning smile.

The month of July came and went so fast. Maggie and I saw each other a little more because I was around the office. We were both less home sick because we were so busy. It was almost like going to college. We were doing and learning things we had never been exposed to prior to this summer.

Our work for the month of August was going to be only three weeks for us as we were going to spend the last week at home with our families. Maggie spent her time learning how to make food healthier. How to use leftovers to create good meals. How to use spices to make plain food special and more desirable.

My last few weeks of education were spent with the ranch veterinarian and the breeding specialists. I received a crash course in how selecting certain individuals in a herd and using them during breeding, could introduce desired traits in future offspring. I sometimes felt that I was working as a doctor in a hospital. The ranch took breeding very seriously and the proof was in their cattle and the demand for their animals. By the time my three weeks were up, I had an arm load of binders that were filled with information.

A few days before we were to leave the ranch and go home to our families, Andrea stopped by our cabins and let us know that Rex and Abby would like for us to spend our last evening with them at their home.

The last day came, and Maggie said goodbye to all the people she had worked with during the summer. Brad let me go after lunch so that I could have time and go around the ranch and thank everyone for putting up with me for the summer.

Maggie and I agreed to meet after she spent some time with Precious. We only had about an hour to get ready to go to see Rex and Abby. We did not have much time to talk about anything. Our work was done for the summer and we just sat together and held hands. It had been a whirlwind of learning in a very short time.

The truck arrived to take us to dinner. I think that my body was just collapsing. I was beginning to relax for the first time in weeks. Maggie looked exhausted too. We had survived and we were together, and we were going home in the morning. How could it be any better?

Wen arrived at their home, Rex and Abby were waiting outside. We got what was becoming an expected hug from each of them. I do not know about Maggie, but all the tension was draining out of me, and I was starting to relax.

I tried to read Maggie's body language, but she was busy talking with Abby and it was difficult to tell her feelings.

We sat and made some small talk before Abby said that dinner was ready. We went to the big table and seated ourselves. As usual Rex prayed before dinner. We had a wonderful meal of steak, squash, salad and a wonderful cake for dessert.

When we finished up, Abby said we should go to the den and relax. Rex asked Maggie how her horse was doing. That was the wrong question to ask her. She could have gone on for an hour. Rex just listened and never stopped smiling. Maggie eventually realized she was doing all the talking and said she was sorry. Abby told her that there was no apology needed, that it was wonderful how she loved her little horse.

Rex looked at me and asked how my roping was doing. I guess he had heard that I needed some more lessons. He was smiling when he asked me, so I knew he was giving me a friendly poke. I told him that during this summer that I had learned more than any three-month period in my life. That seemed to make him happy.

Abby asked Maggie about her job in the Chuck Wagon. Maggie expressed that she also learned far more than she had expected. We both related some of the favorite memories we would take with us as we left the ranch.

Rex looked a little more serious as he began to speak. "I wanted to explain my thinking about what you have been asked to do this summer. I know that you missed these few months with your families. Each of you studied hard in school this past year and your grades reflect that effort."

"I wanted to see if you could handle the pressure of even more responsibility by spending the summer here. It may not have been apparent to you, but I have carefully followed your journey this entire summer. There were reports created by the folks who supervised your work on the ranch. Abby and I wanted so badly to have you spend more time with us. That would have interfered with this test of your strength. So, we had to watch you at a distance."

"I want to tell you that you have exceeded our expectations. I am proud of both of you. As you leave for home, please express my thanks to your families for allowing me to have you spend so much time on the ranch."

"This is a bit early, but I want to tell you that next summer I would like you to come straight from school to the ranch. We would like for you to spend one month here and then two months with your family. There are several things we would like you to become familiar with before you go home, the day after our fourth of July celebration."

"When you return to your home, we would ask you to work in our businesses for the rest of July and then to have the entire month of August free to do as you please. "

"When you get home, I think you will find a lot of changes that have happened in the last couple of months. I hope you find them positive."

"I want you to know that if you ever feel that your life is being drawn in a different direction than it is now, please talk with us. We only want the best for you and your happiness is our main priority."

As he was speaking, Abby seemed to be quite emotional. A time or two, I noticed that she looked away and rubbed her face. It was odd to me that these two people who had been strangers not so long ago, were acting like our family members. I thought how charmed my life is to have Grandma and Grampa and now Abby and Rex in my life and all of them looking out for my future. How blessed I am.

We talked about small things for a couple more hours

before Rex said that he was being selfish and should let us get to bed. As we were preparing to leave, he went to the other room and returned with some folders and two envelopes.

He handed each of us a folder and an envelope. "These are your class schedules for the fall semester. The envelope contains your wages for the summer. Thank you for living up to our expectations."

Rex told us the plane would be ready at 10am. We took turns hugging each other. Abby kissed me on the cheek. Maggie and I walked out to the truck. Rex and Abby were waving as we drove away.

Neither Maggie nor I had anything to say as we drove back to our cabins. Saying goodbye to Rex and Abby was becoming harder with each parting. Part of me was sad about leaving them and the other part was excited to be going home to see Grandma and Grandpa. I am sure Maggie felt the same.

# CHAPTER TWENTY-FOUR

Maggie and I were up early. We had breakfast and then drove out to the pasture to see Precious and her mom. Maggie climbed through the fence, and that little horse and her mother came right to her. I stayed outside the fence and let her have this time to herself.

She hugged and kissed her horse at least a hundred times. It was a lesson in how to express your love for another living thing. When she finally let go and was trying to climb through the fence, Precious was trying to come with her. It was a tender moment. Maggie softly told her no and pushed her away.

When Maggie finally looked at me. She had tears in her eyes. I smiled as she wrapped her arms around me and started to cry a little more. We got in the truck and drove back to our cabins to get our suitcases.

When we decided that it was time to go meet the plane, we took one last look at the little homes we had lived in for the summer.

Two trucks arrived at the cabins. One was our ride to the plane the other was full of the friends we had made that summer. They drove out to the plane with us, and we had a round of hugs and well wishing. It was a wonderful send off. The trip home was very nice. The closer we got the more excited we became.

Maggie and I decided that we should maximize the time with our families as we would only be there for a week. We knew we would be together much more at school than we had been while working at the ranch. We were sure looking forward to that alone time in a week or so.

As the plane neared the airstrip on our ranch, we could see our families standing at one end and waving at the plane. We landed and taxied over to where they were standing. Maggie and I thanked our pilots, and we stepped out of the plane into the arms of our waiting family members.

There was lots of hugging and holding. It was wonderful to be home and see all of them. After a few minutes, Maggie and her family left for town , and Grandpa and I left to go see the Amos property.

It had been two months since I left and there were changes everywhere. The house had new paint. Grandpa had obviously relented because there was a little garden by the house.

We entered the house, and I headed for my bedroom to unload my things. When I had finished, it was time to tell my family all that I could remember about my summer.

They let me talk for more than an hour about life on the ranch. When I had nothing else to say, they both had questions for me. We talked all afternoon and into dinner. It was wonderful to be home and sharing dinner with the two most loving people in my life.

The next few days were wonderful, just spending time together and seeing all the changes on both ranches. There were more buildings being constructed. The ranch housing was nearing completion. I met many new workers. Each of the businesses, had hired additional workers. They were no longer just little family projects. The combined activity on both ranches and their companies, made it a huge process.

The holding pens were filled with a much larger number of cattle than when I left eight weeks before. I saw at least a hundred horses and two new hay barns, where a corn field used to be.

It was becoming unrecognizable from when Mr. Amos owned the property. There were lots of refrigerated trucks loading up meat at the processing plant. The company now owned a whole fleet of forklifts that were needed to service all the outbound and inbound trucks.

Grandma and Grandpa looked wonderful. They did not have the stress lines on their faces that I had seen for most of my life. They looked healthy and seemed to be enjoying life. It made me happy to just watch them living their lives without worry. Rex had made such an impact on each of our lives. We were truly blessed. I owed him for everything he was doing for my family.

The time had gone so quickly. I saw Maggie only once that week when she and her parents were touring the properties. Grandma invited them over for dinner the night before Maggie and I were leaving for school.

We had the usual feast that Grandma had prepared. Lots of wonderful conversation by our families, who had been so influenced by Rex and his vision of the future.

The evening went to quickly and we planned to meet again at the plane in the morning. Grandpa and Grandma stayed up late with me. We talked about the next school year and made some tentative plans for Christmas vacation. Before we knew it, the clock was showing eleven o'clock.

We hugged and went to bed. In the morning it was our last time for several months to share a meal. Grandpa prayed for our safe trip and God's blessing. We had talked ourselves out the night before, so we just enjoyed being together.

A little before the plane was to land, we drove Grandpa's truck out to the air strip. Maggie and her family were already there and were standing by their car. We got out and within five minutes the plane came into sight. It landed and came to a stop near us. The pilot opened the door, and we carried our suitcases up the stairs.

We ran back out for one more goodbye. There were no tears this time, just lots of hugs and kisses. Maggie and I got back on the plane and a couple of minutes later the plane had turned around and we were heading down the strip. Our families were waving as we left the ground.

The flight went by quickly and we were delivered to the airport near our school. We took a cab to the campus and unloaded our things in our respective dorm rooms. I told Maggie we should find our truck and go to town for a meal.

We took our folders Rex had given us and while we were eating, we looked over the schedule of the classes we would be taking this year. This being our second year of school, many of the things we experienced last year were not going to be a factor. We knew the campus layout. The same counselors we had the year before would be helping us again. It was going to be a much less stressful experience.

When dinner was over, Maggie wanted to go see a movie. That is how we spent our last evening of freedom. It was a fun night, and we cut it a little short so there would be time to prepare for classes the next day. Our schedules showed that this year we would not share any classes. It was shaping up that the weekends would once again be our only real time to be together.

The first week was good. The new classes were going to be interesting. We saw our friends from the ranch and a few other folks we had met the year before. None of my classes had any homework the first week. The same was true for Maggie. We planned a fun weekend with Darren and Janie from the ranch.

Janie had a classmate who lived in town. Her family owned a boat and had invited us all to go water skiing with them on Saturday. The day was wonderful. We had a huge picnic, lots of time on the boat, and a campfire when it got dark. It was one of the best weekends we had experienced since starting school, the year before.

The weeks passed by filled with classes, homework, and weekend time together. Andrea checked on us periodically. We were already making plans for Thanksgiving and Christmas. Rex had left those holidays open for us to do as we wished.

Maggie and I traveled home for one week in November and one week in December. We enjoyed the time with our families and were able to catch them up on everything happening at school.

The second semester was a grind with no time off until the Easter break. Rex had contacted us and asked if we would be ok with visiting the ranch during spring break. Maggie and I talked it over and decided since we were going to have six weeks at home during the coming summer, we would go see Rex and Abby.

I called home and talked with Grandma because Grandpa was gone to town. I explained what Rex wanted and she encouraged me to follow my heart. The night before spring break started Maggie and I drove to the ranch. We got in quite late. Both of our cabins were ready. We were very tired and parted ways with some kisses.

## CHAPTER TWENTY-FIVE

When I got up the next morning, there was a note that had been pushed under my door. I opened it. Maggie wrote the note which said that she was gone to see Precious and would see me there.

I got ready and found a truck to drive. As I approached the pasture, I could see Maggie with Precious and her mother. Not wanting to interrupt the moment, I stopped the truck and turned off the engine. It must have been thirty minutes later that she saw the truck and waved at me.

I drove over and got out. The sun felt good and Maggie's face was radiating her love for that horse. It would be impossible for me to not love this girl who had so much to give. Maggie did not know it, but I had purchased my own camera with my money from the summer with Rex. While Maggie was playing with her horse, I had taken several pictures.

When Maggie was ready, we drove to the Chuck Wagon for breakfast. After a nice meal and hearing Maggie talk non-stop about Precious, we walked over to the office to see Andrea. She gave us both a hug and told us that Rex and Abby would like us to come to dinner at their home that evening.

Maggie wanted to go swimming, so we went to the pool for most of the day, leaving with just enough time to get ready for dinner. It had been a very nice day, free of stress, books, and tests.

As Maggie was ready to knock on their door, Abby opened it and stood there with a smile. "Welcome strangers!" The usual round of hugs and kisses followed and then it was repeated when Rex showed up.

Unlike some of our other visits, we did not rush into

dinner. Everyone sat in the den and talked for more than an hour before Rex said we should probably eat something. We sat down in the dining room, Rex prayed, and then we spent another hour talking and eating.

As dinner was winding down Abby said, "We would like you to spend a few days with us at our home." That was unexpected. She went on to say that they wanted to spend time with us without the ranch people wondering what was happening. She asked us to go pickup our things from the cabins and return for some more conversation before calling it a night.

We drove to the cabins, packed up our things and returned to their home. Abby showed us to our private guest quarters. Each of our rooms was beyond amazing. There would be no problem enjoying our stay in this house.

When we were unpacked, Maggie and I returned to the den. Rex asked us if we were up for an adventure in the morning. We both assured him that we were ready. He told us that we would be flying to Humbolt, Kansas and looking at some bulls that were for sale.

Rex told us we would leave early, and we could have a breakfast on the plane. Abby said she would give us a wakeup call in the morning. Rex said. "See you in the morning, bring enough clothes in case we decide to spend the night." He and Abby left us alone in the den.

It was like some sort of fairy tale. Maggie and I were alone in this beautiful room. We were going on a plane trip in the morning with our wonderful friends. Life was good.

We cuddled up on the sofa and held each other. Maggie was falling asleep, so I kissed her and said we should get some sleep. We walked to our rooms, had some more kisses before we parted.

I was dead asleep when Abby knocked on my door and

said, "Thirty-minute call". I answered "ok" and headed for the shower. By the time I checked in on Maggie, she was already gone. The three of them were waiting on me with their little bag of clothes. I had to run back to my room because I had forgotten mine.

We drove to the airstrip. Rex had a new plane that we had not seen before. It was easy to see that he was looking for some reaction from us. It did not have to wait long to hear how impressed we were at his newest purchase.

The flight was like a dream. The plane was very large inside, and the seats were more than comfortable. The noise was far less than what we had experienced in the smaller planes. Once we were at the desired altitude, Abby showed Maggie where our breakfast was stored, and they prepared the meal for us. Rex seemed so happy. Maybe it was because he was going to enjoy looking at some new bulls or maybe it was because Maggie and I were traveling with him. Either way, he was enjoying the trip.

The plane landed in Topeka, and we were picked up by some people from the ranch we were going to visit. The countryside was very different from what I was used to seeing. It was flat and you could see a long way in any direction.

It did not take long to reach the ranch headquarters. We were met by members of the Marshall family. Their people had owned the land for more than 100 years.

After some introductions, we were shown to our lodging. It consisted of a large two-story home, about a quarter mile for the headquarters building. We were told that we should feel free to look around and dinner would be at five o'clock.

We settled into the house, and Rex told us that we could go explore on our own. He asked Maggie and Abby if they wanted to go along. Abby said that she and Maggie had their own plans. Rex and I left alone. It was his goal to see as much of the operation as was possible before we had to meet for dinner.

Adam Marshall, the father of the family, had left us a truck to use during our visit. Rex and I took off and did our own exploring for several hours. The ranch was clean and well maintained. The pastures were weed free and the cattle we saw looked extremely healthy.

I asked Rex what breed of bulls we were going to be looking to buy. He told me that the Marshall Ranch had some of the best Hereford and Charolais bulls that were available. He was particularly interested in the Charolais because of their high-quality carcasses and feed conversion.

After looking around for some time, we saw a pasture of nothing but Charolais cattle. They were too far away to see if they were bulls. Rex drove us in their direction, and we were able to see that they were cow and calf pairs. I was impressed by the animals. They were large and well-muscled. Light in color with very broad heads.

Rex said that the studies he had read showed that these cows produced some of the heaviest calves. He asked me what I had learned at school about this breed. I told him that we had briefly discussed it, but we had not spent much time on any in depth study of the breed.

He told me that we needed to learn all we could about the Charolais cattle because he felt it could be a way for us to produce a superior product by crossing our cattle with bulls from this ranch.

The rest of our day was just driving and talking. I spent more time alone with Rex that day than I had at any time in the past. He was a man on a mission to build a cattle empire. We got back in time to clean up for dinner.

The Marshalls were a very nice family. It was a very nice

evening. There was no conversation about bulls. That was going to wait until morning. We learned a lot about their family roots, and they asked Rex about his family. Rex agreed that he and I would meet Adam in the morning and spend the day looking at bulls and asking questions.

The amount of information I learned the next day was like a college class about Charolais cattle. I wished that I could have taken notes. Rex seemed to be absorbing every detail that Adam told us about his bulls. We looked at more than fifty individual bulls. When Rex showed any interest in a particular bull, I used my camera to take pictures. Adam took us back to headquarters and pulled up an information file on each bull that Rex had been interested in buying.

After more than two hours looking at files, Rex told Adam that we would go outside and talk before making any decision. To my surprise, Rex wanted my input as to which bulls I felt we should consider. I did have a list that I made during our time looking at files. So, I showed it to him.

He had a list in his head. I do not know how he was able to do it. He asked me to write down five bulls which were listed by their ear tag number. He wanted me to rank them from first choice to fifth choice. It took me a few minutes because I was trying to remember the details on each of the bulls I had chosen.

When I had finished, he took my list and looked it over for a few minutes. He handed it back to me with a smile and said, "Well Tommy we agree on four of the five and we also agree on the ranking." He patted me on the back and said, "let's go inside and you can tell Adam which four we want."

That is how the Walker Ranch purchased the bulls that would change how our cattle were bred. Rex made a deal for the sale price and set a date when our truck would arrive and pick them up.

We went to the house and told the girls that we had purchased some bulls, and we were ready to go home. The pilot had the plane ready when we arrived back in Topeka. We had a nice flight home. Our truck was still parked by the airstrip, so we loaded up and drove back to the house. Rex said, "We need to celebrate our successful trip. Let's have a barbeque tonight."

It was a wonderful evening. Rex cooked the steaks while the women made salad, mashed potatoes and squash. We were all hungry because we had skipped lunch. It was a wonderful evening together. I do not know about Maggie, but I was feeling that Rex was helping me grow into a new role.

As we were going off to bed, Maggie asked if we could talk for a while before we went to our rooms. It was a lot more than a while. She went on and on about her alone time with Abby. By the time she was done talking, it was clear that Abby was doing with Maggie exactly what Rex had done with me. We were being groomed for something far beyond what we could see as sophomores in college.

Maggie was bonding more with Abby, and it was wonderful to watch. Other than her mom and dad, I had never seen Maggie have such a close relationship as she was obviously having with Abby.

Maggie wanted to hear all about my day with Rex. I told her that I would save it for the next day. We were both tired and she was happy to wait to hear my story.

At breakfast, Rex told me that he would send someone to town to have my pictures of the bulls developed and enlarged. He wanted to have the larger copies framed and hung in his office. This man seemed to know how important these new additions to the ranch were going to be.

Abby announced that she and Maggie had some things to do and that we were on our own until dinner time. Rex wanted to go talk with the ranch veterinarian and the crew that handled the artificial insemination. His mind must never stop.

By the end of the day Rex knew how many cows the four bulls could service each month. He used those numbers to do some long-term planning. I was included in all the conversations and planning. It was making my classes at school seem much more relevant.

The days went far too quickly and our trip back to school was just one day away. Our last evening with Rex and Abby was the culmination of so many things that had happened on this visit.

It was somehow like spending an evening with your family. I do not think that Maggie or I felt that our relationship with the Walkers was a boss and employee situation. It was much deeper than that. The four of us shared a trust and a mutual respect. My life had been so small compared to the Walker empire. It had been just Grandpa, Grandma and me for most of my life.

Our little farm and our animals had been my world. I had never been out of the state until Rex came along. Things were changing rapidly. Maggie and I were going to school in another state, working on the Walker Ranch, and having our future planned by Rex and Abby. It was a major change for two small town kids.

I got up early the next day. Before we had gone to bed, Maggie had told Abby that she would like to spend some time with Precious before we left for school. She and Abby were already gone, so Rex and I had breakfast together.

He told me that he would like to work up a chart that showed the four new bulls and the expected rate of using them to raise a new generation of our cattle crossed with those new bulls. I asked him a few questions about all the information he would like to see on the chart. We spent an hour deciding the final features of my new assignment.

When we had finished, he said, "Tommy, this could be a major change in the right direction of this ranch. I wanted you to be part of it from the very beginning." I told him that it was a very exciting plan and that it would be a new adventure for both of us. I guess he liked that because he said, "Here is to a new adventure." Then he held up his glass of milk and we had a toast.

We had just finished when the girls came in the door. We said our goodbyes and had some hugs and kisses. Rex said he would be in touch. Maggie and I drove back to the cabins, loaded up our things and left for the school.

## CHAPTER TWENTY-SIX

The trip back to school went very fast. Maggie told me all about seeing Precious and the things she and Abby talked about when they were together. I gave her a summary of the things Rex and I had planned. It would be safe to say that these two kids were very happy with our lives. We had a future, loving parents, Rex and Abby, and each other.

The rest of the second semester went along well. Maggie had been given a few projects to work on for Abby, and I was well into my projections and the chart Rex wanted.

Maggie and I made it a priority to spend some weekend time together. During the year we had been making a list of interesting things to see on our days off. We enjoyed seeing as many of them as we had time.

The last week of school was the usual cramming for finals, followed by day after day of taking exams. Somehow, we survived the week. On Saturday morning we loaded up our truck and headed toward home.

Maggie had completed all the things Abby had asked of her and I had submitted the chart Rex wanted. We were free for 6 weeks before we needed to report to Walker Ranch. The trip home was uneventful, and we arrived at Maggie's home early in the evening. We had a warm welcome from her parents. I helped unload her things and headed to my home. Stephanie told me she would call Grandma and tell her I would be there in a few minutes.

They were both sitting on the front porch when I drove into the yard. It was a wonderful sight to see. The two people who were my family, sitting and waiting for me. We did not even unload my things that evening. All the time before bed was spent talking and sharing. I did not realize how much that I had missed them.

The next morning, while we ate the breakfast, Grandpa announced that if I was up to it, he wanted to take the trip to Tahlequah that he had been planning for so many years. It was fine with me, and we made plans to leave the next day. I could not remember the last time that I had seen him so excited. It was going to be special. Just the three of us going to see Grandpa's roots.

We did not need to leave early that morning. Our destination was less than a two-hour drive. It seems almost impossible that we could live such a short distance from this town and yet have never visited it in my lifetime.

We checked into our motel, and Grandpa showed us a list of things he wanted to see. Grandma did not discourage him. He had a long list, and she was willing to stay as long as it took to visit every place on his list.

Neither Grandma nor I had any Indian blood as far as we knew. That made this trip all about Grandpa and we were both happy to share this time with him. That afternoon we drove out to Park Hill and visited the Ross Cemetery. John Ross had been the principal chief of the Cherokee People for several decades before he died in 1866. Grandpa moved his hand over the surface of the gravestone. No one said anything. It was a moment in time where one Cherokee descendant honored another.

There were several other important individuals to Cherokee history who were also buried in this place. We visited them all before we headed back to Tahlequah. Grandpa finally said, " It seems like such a humble place where the leaders of our people, were laid to rest." That was all he ever said about the visit to this spot.

We stopped at a little dinner that had a sign saying, "Traditional Cherokee Food". Grandpa wanted to stop there for our late lunch. Grandpa recommended that we try some fry bread, beans, and selu gadu (corn bread). It was very good but lacked the variety that we had when Grandma fixed lunch. Grandpa seemed to enjoy every morsel. Grandma went along with everything, but I could tell she was not impressed.

Our last stop that day was the Cherokee National Prison which was built in 1875. We had dinner at a regular restaurant. After arriving back at our motel, Grandpa talked about his Cherokee relatives. He told me stories that I had never heard from him before that night. Maybe it was this place or his age. It just seemed that many things about that part of him that was Cherokee were suddenly coming out. It was all so new to me. Other than a few words in Cherokee and one or two stories, he had never said much about being part Indian.

We spent two more days in Tahlequah before heading home. We visited two museums and several other sites that were important in Cherokee history. It occurred to me that this may have been the final thing on Grandpa's bucket list. As we drove home there was a certain peace that came over him.

Those few days showed me a side of my grandpa that helped create a complete picture of this man that I loved so much. I was proud of him and his heritage. Thanks to that trip, I learned a lot more about the Cherokee People and their struggle to survive. I believe that the total experiences on that trip, helped me become a humble man for the rest of my life.

My Grandpa never lived a day of his life where he did not thank the Unetlanvhi (creator) for all the blessings that had been bestowed on our family. I tried for the rest of my days to be the type of man that my grandpa had been to all the people he came across in life.

When we got home, Grandma stepped out of the truck and hugged Grandpa and said, "Roy, thank you for taking us on that journey." Then she kissed him. It was a sight that I will never forget. That was the one time I think that I ever saw a tear on his face. It was a type of closure or finish for him. I am forever grateful for the opportunity to have shared that time with him.

The next few days, Grandpa and Grandma enjoyed showing me all the new things on the ranch and at the businesses. Our properties had become a working community. There was new housing, barns, equipment, cattle, horses, workers and even some new signs. But Rex had never taken down the sign Grandpa had put up all those years ago.

There were more tables down at the lake and some playground equipment. Rex had even started stocking fish in the lake so that the workers could enjoy the fishing.

It was taking on the appearance of being a Walker Ranch. There were more cattle and horses than there had been a few months ago. Grandma and Grandpa still worked in the businesses, but it was now down to only three days a week. Andy and Stephanie were on the same schedule. Rex had promised them that it would be down to two days at the beginning of the new year and that he hoped they could retire the following year. I had learned from Grandpa that both entities were highly profitable and that Rex was very pleased.

On the days that our parents were at work, Maggie and I had a chance to be alone. We got to know a lot of the new workers. Maggie enjoyed spending time with the horse trainers. She was learning a lot more about how to get a horse to the place where you could ride him or her. They seemed to enjoy teaching her as much as she liked learning.

I was becoming familiar with the feed lots and the rations the cattle were fed. While Maggie was learning about training horses, I spent my time listening to the workers who handled our cattle.

Rex was a brilliant man. He was using our off time as additional training time, without us even realizing it. One afternoon, I had a chance to be alone with Grandma. I asked her about the trip to Tahlequah with Grandpa and me. She told me that she would not change a thing about that trip. "Roy loved every minute of that time. I believe he will die a happy man." I asked her if there was something she would like to do so she could die a happy woman.

"Tommy, I have been giving that a lot of thought. Seeing the Cherokee people and knowing how much Roy loves them has made me want to do something for them." I asked her if she had come up with anything and she replied, "Yes, I have. I would like to create a healthy cookbook and give it away to them for free." I thought about it for a moment and told her that I thought it was a wonderful plan.

While we were eating dinner that night, I brought up the idea to Grandpa. He was his usual self and had to think about it for a little while. "Ida, I think it is a wonderful idea. How can I help you with your project?"

And that is how it all started. Grandma shared her idea with Stephanie, and she told Andy. He contacted some government people he had become friends with during the war. Those people were able contact the right people about getting a government grant to supply a healthy cookbook to the Cherokee people.

Before Grandma even had her book complete, all the necessary papers had arrived and been signed and returned to the government offices in Washington DC.

Stephanie helped Grandma for many weeks. They picked

recipes and Grandma would tell Stephanie the ingredients and the quantities. After a few months, the book was ready. Andy did the research and found a publisher that was willing to discount the normal cost because it was a nonprofit project.

As the book was nearing completion, Grandpa suggested that it should have a title he picked out. That is how it came to be called "GRANDMA'S HEALTHY COOKBOOK". It was published in the fall of 1952, and Grandpa had the honor of personally delivering the first shipment of books to Tahlequah.

Representatives of the tribe accepted the books, and they were distributed to individual Cherokee families by the tribal government. It was a wonderful thing to see my family coming together over a common goal of helping Cherokee people. Grandma and Grandpa were doing a wonderful thing in the later part of their lives.

Our six weeks came to an end much quicker than we would have liked. Maggie and I packed up our things the night before we left. I picked her up early the next morning and we were off to the Walker Ranch.

We were very happy to have had the time at home, and it made leaving for the next few months, easier to accept. The long drive to the ranch was the first time all summer that we had an entire day to ourselves. We talked about Grandma's book and how badly Maggie wanted to see Precious.

We were just a few weeks from starting our third year of college. By next spring both of us would turn twenty years old. I guess we were leaving our kid years behind. Maggie was becoming more beautiful each year. She inspired me with how much love she had inside. I had never seen her be unkind to any other human being. It would be difficult for me to even think of a future without her.

We arrived at the ranch very late that night. As usual our cabins were ready. I helped Maggie unpack and went next door and crashed on my bed without even unpacking. There was a note on my dresser to meet Rex at his office at nine the next morning. Fortunately, I had awakened early enough to see it and let Maggie know. We went to the Chuck Wagon for breakfast and then met Rex at his office.

He seemed very happy to see us, and we had our hugs before starting to talk any business. Rex told us that Andrea was doing the finishing touches to our fall schedules. They should be ready for us to see in a week or two.

"As much as I would love for you to stay with us at the house, there are some opportunities that have opened up and we need to take advantage of them. Maggie, I would like you to go to Denver and work with some federal food inspectors for the next thirty days. That will give you some insight into how that process works. You can then help us with our planning, so that we will be compliant with any program we decide to implement."

" Tommy, I want you to go to Chicago and work with a very large commodity trading firm. They can teach you about the futures market and how we can plan when to sell cattle, when to buy and sell grain, and how to hedge our business to protect us from the ups and downs in the market. "

Maggie was the first to reply, "I think that sounds very interesting and I am willing to go." Rex smiled one of his famous smiles and said, "I knew that I could count on you Maggie."

I basically said the same thing Maggie had said, and Rex gave his approving answer. I asked when this was all going to happen. Rex told us the plane would drop me off in

Chicago and then take Maggie on to Denver. He apologized for the short notice and said we could have the following day to ourselves, and the plane would be ready the following morning at nine. He said, "Abby would never forgive me if you did not show up for dinner tonight. I hope you will come." We agreed and then he told us get out of here and go see Precious.

We left his office with our heads in a spin. This guy never stopped thinking. I believed we were going to have a nice working vacation on the ranch for the next few weeks. Boy was I wrong. We found a truck and drove out to see Precious

At first Maggie could not pick her out. She had grown more than Maggie had expected. She stood at the pasture fence and Precious did the work for her. She came running over to Maggie. It did not take Maggie very long to climb the fence and embrace her little horse who was not so little anymore.

I had my camera and was snapping pictures as fast as possible. It was a touching moment. I was trying my best to capture on film the bond that they shared. To watch her interact with Precious just made me love her more.

We ended up back at our cabins with a couple of hours to spare before needing to get ready for dinner. It was a warm day, so we stayed inside and used our time to talk about the plans Rex had made for us.

Life was never boring around Rex. How in the world did he make all these plans and somehow get the cooperation of people in these industries to go along with his ideas? You could only admire his ability to get things done.

We spent some time repacking after talking for an hour or so. Maggie ran over to the Chuck Wagon and got two apples. She wanted to visit Precious and her mom before we went to dinner. I took some more pictures as she gave both horses a few bites.

Dinner was as special as every other time we had spent an evening with Rex and Abby. The meal was delicious and our time with them was wonderful. Abby was concerned that we might feel that we were being pushed too hard by Rex. Maggie assured her that it was just one more adventure for us. That seemed to make her feel better.

When we parted for the evening, Rex told us he would not be able to see us off at the plane in the morning. He and Abby were leaving very early to meet with some potential investors in Dallas.

We had our hugs and some kisses as we left them that evening. When we got back to the cabins, Maggie wanted to call home and tell her parents about what we would be doing for the next month. While she was calling, I did the same. I do not believe that anything Rex did would surprise either of our families. They were supportive of our decision to go along with the plans that had been made for us.

We met the plane, and it was only a few minutes before we were airborne. The flight was several hours long. I shared with Maggie that I was a little uneasy about my assignment. She kissed me and said, "Gule, you will do fine." Her confidence in me was wonderful but it did not make it any easier for me.

We agreed to call each other every night while we were apart. As usual we each had a folder with instructions for what to do when we landed. We were nearing Chicago, so I used the time to review my instructions. A driver would meet me in the terminal. He would take me to my hotel. That evening, I was to have dinner with some people from the company of Wilson, Abbott & Marks.

The plane was only on the ground long enough to drop me off and refuel. This month would be the longest time that Maggie and I had ever been apart since her family toured the National Parks, many years before.

I did not even get to see the plane take off as my driver was waiting inside the terminal. He drove me to my hotel, and I had a couple of hours before dinner.

I decided to call home again. Grandpa answered and told me Grandma was over at the plant. We talked for some time before he said " Gule, no one knows where our path will lead. The Creator keeps that knowledge to Himself. I believe that if we could see the future, we would not be comfortable with what we might see. Trust Him with your path and it will turn out fine." It was a good conversation, and I was a little less apprehensive about the next month.

## CHAPTER TWERNTY-SEVEN

The evening was nothing like I had imagined. The driver picked me up and delivered me to a downtown restaurant where I met three of the directors of the firm where I would be working. We shook hands and they introduced me to themselves. I felt a little underdressed for the occasion. No one seemed to mind.

We ordered dinner and there were no prices on the menu. I had never seen anything like that in my entire life. I really felt like a smalltown kid eating dinner with movers and shakers of the world commodity market.

They were very kind to me. They seemed most interested in my relationship with Rex and the Walker Ranch. I got the impression that it was a personal favor to Rex when he asked them to give me a crash course in commodity trading. I guess they were looking for whatever it was that made me special to Rex.

I thought about it later that night. Why would a huge company send three of their top executives to eat dinner with a soon to be junior in college? It was all about Rex. He must have been a very important client of their company.

It was a very special dinner. We had some things that I had never tasted before. Some of them had names that I could not even pronounce.

The evening ended with each of them wishing me the best while working at their company. The driver delivered me to my hotel and handed me a folder of instructions. He told me he would pick me up at six the next morning. I asked him why so early and he told me the commodity markets open early each morning.

It took me a couple of hours to finally fall asleep. My mind was reliving the events of the last week. My last thought before falling into a restless sleep, was that I hoped everything was going well for Maggie.

When my alarm went off, I felt as though I had only been asleep for an hour. I was very tired and not looking forward to having to be alert for the next several hours. I wanted to call Maggie but there was not enough time. I showered and met the driver at the curb in front of the hotel.

He delivered me to the employee entrance of the firm and gave me instructions on how to find my contact. My teacher for the next weeks was on the third floor, room 312. His name was Robert Miles. I knocked on his door, and he opened it with a big smile. I started to tell him my name, and he said, "I know who you are, come on in Tommy." We shook hands and he started talking about the job. I guess he had been briefed about me, and it was down to business.

"Tommy, the commodity business is brutal, but it is the most exciting game in town. Set your mind to the task at hand and in the next four weeks you can learn some life changing things about food and financials. Most importantly you will come to realize how a small group of
importantly you will come to realize how a small group of
individuals control the entire world market."

He explained that for the rest of the week, he would be giving me classroom instruction on the basic concepts that I needed to understand before we could advance to the more important subjects. He stood up and led me to a classroom just down the hall from his office.

After offering me a cup of coffee, he started my education. Every day was the same for the rest of the week. We met in the classroom, our lunch was brought to us, and we never stopped the fast-paced learning until quitting time at four o'clock.

When I got back to my hotel after my first day, I fell onto

my bed and did not wake up for two hours. There was a light flashing on the phone by the bed. I called the front desk to see what message they had for me.

They had a message from Maggie. She left her room number and the hotel where I could call her. We had a long conversation about our first day of our month-long job. We took turns telling each other every detail.

Maggie had an older lady who was her trainer. She knew everything about the food industry and the role of the inspectors. The trainer was named Hilda Sorren. Maggie was to have five days in a classroom before moving to other duties.

We agreed to try and call each evening to discuss the work that day. With only a couple of exceptions, we were able to do that for the entire month.

My week was spent learning the basics about supply and demand, why futures contracts were used by traders around the world, and how the markets controlled the availability of needed items.

Maggie learned why inspectors were needed, how many industries had inspectors and how the food supply was protected from disease and poor quality.

Each day I gained new knowledge that was built on the foundation of the previous day's learning. Both Maggie and I were being exposed to things that few individuals understood or were even aware that they existed. It was beyond what two teenagers would normally be expected to understand.

By the end of the week, we were both on overload. We spoke by phone for several hours on the first Saturday morning. We missed each other. This was harder than any college classroom experience that either of us had ever been exposed to in our short lives.

Maggie told me she had picked up some flyers in the hotel lobby and was going to do some exploring on Sunday. When we were done talking , she was going to call home. I agreed with her and I call home after Maggie and I finished our conversation.

After talking with Grandpa and Grandma, I went down to the lobby and asked about any fun things to do in town. The desk person told me about a nearby museum and a movie theater. I visited them both before going to a very nice place for an early dinner.

Maggie and I talked again on Sunday morning and then we both spent the rest of the day going over our notes from the first week of work.

As it turned out, both Maggie and I had another full week of classroom instruction. It was less stressful because we were building on subjects we had already learned. Neither of us could have asked for a better teacher. They knew their subjects and were happy to work with a couple of young people who were eager to learn.

On the second Friday of my training, Robert told me that I was to meet with a few of the directors before the end of the day. He told me not to worry. It was just a formality before going onto the next phase of training.

About an hour before the day ended, an assistant of one of the directors took me up to the top floor. In a very elegant meeting room, three of the directors were waiting for me.

The president of the board of directors was named Mr. Hamilton. He greeted me and pointed toward a chair where he wanted me to sit. When I sat down, he introduced the other directors and then began to address me. "Tommy, over my years on this board I have had this same meeting with many individuals, who have gone through a similar training process. This folder before me is a record from Robert Miles on your current progress. I am happy to tell you that I am very impressed by what he has said in his report. I have discussed what I am going to tell you with the other members here today. We are all in agreement with what I am going to say."

"The next two weeks of normal training is going to be set aside, and our company is going to fast track you into a much broader education than we usually offer new employees. It is very apparent that Rex Walker saw something special about you and that is why he sent you to us. You have made quite an impression on Robert Miles, and he has convinced us to invest more training in you."

"We have only one request. If for any reason you do not go forward with the plans Rex has made for you, will you entertain a job offer from our company?"

This was the first opportunity where I was given a chance to respond to anything said so far in the meeting. I took Grandpa's advice and considered my response before speaking.

"Gentlemen, your confidence in me is greatly appreciated. Your offer is the thing dreams are made of ,and I do not take it lightly. Rex Walker is more than an employer to me. What he has done for my family can never be repaid. I made him a promise and the only reason I would not go to work for him would be if he wanted me to do something else. Should that happen, I would seek his blessing in accepting a job with your company."

Mr. Hamilton smiled and said he did not expect anything less than my answer. Then he said, "Rex Walker is a lucky man." We shook hands all around and everyone wished me well in my next two weeks of training.

That evening, I called Maggie and told her about the

meeting and what had been said. She said she was proud of me and was not surprised that the firm had offered me a job. She told me that her next two weeks were going to be in the field where she would be checking compliance with another agent. She was excited.

As we were hanging up, I managed to say for the first time, "I love you with my whole heart Maggie Scott." There was a moment of silence before she said, "I have been waiting for you to tell me that for a long time. I love you as well Tommy Meadors. You make my life complete."

Our third year of college was about to start. We pretty much knew our future jobs and now we knew we had each other. I suppose we always knew it but now it was out in the open. The truth is I had loved Maggie pretty much my whole life.

## CHAPTER TWENTY-EIGHT

We both spent the next two weeks learning more than we could have dreamed. The time went quickly and each of us was given a going away party. Maggie had been offered a job working for the government. She gave them a similar speech to the one I had given the firm.

The following day, the plane picked me up and we flew to Denver where Maggie boarded and together, we flew back to the ranch. It seemed like I had not seen her for a year. She was so beautiful. We snuggled together for the flight back to Texas.

Rex met us at the airstrip. We had our hugs, and he took us back to his home where Abbie was waiting. It was like a family reunion. Abbie told us that we were spending the night with them. She asked us to get unpacked and clean up for an early dinner where she wanted to hear all about our last month.

During dinner and for several hours after, we related our activities for the last few weeks. These two people had become family members to both Maggie and me. They appeared to take great pride in our learning process.

The next morning Rex and Abby had breakfast ready when we came out of our rooms. While we were eating, I asked Rex what the plan was going to be for the following summer. He spent a little time in thought as he composed his answer. "Tommy, there are a lot of things happening in the world that will change the future for generations. Many of those changes will be influencing our lives here in Texas. I will be giving all our futures a lot of thought over the next few months.

My answer is that I do not have a final plan yet. I want you both to concentrate on your studies and plan on completing your junior year of college. We will not make any final plans until later this year or early next year. Plan on spending Christmas with your families."

That was his answer. It created more questions for both Maggie and me. This was a man who was always planning into the future. He was meticulous with details. To have him admit that he was still unsure of what was to happen next summer, created a little concern for us.

We spent our last remaining days on the ranch, with lots of time for Precious and a few adventures for ourselves. We had one more meal with Rex and Abby the day before we drove back to school.

The trip was fun because we were alone. The scenery was very familiar by now as we had made the trip many times. We spent our time enjoying this time together.
As usual we had our folder containing the plans for our third year of classes.

We arrived back at campus with plenty of time to start school on Monday. The routine was old nature to us by now and we headed into the fall semester as veterans. The weeks went by quickly and the weather changed from summer to fall and soon entered the early stages of winter.

Maggie and I spent every weekend together. We explored more of Texas and even some nearby attractions in other states. But we mostly just allowed our feelings for each other to grow stronger.

It was our tradition to call home each weekend and share our lives with our families. There was not much news on either end, but it was something special which kept us connected.

The semester exams came to an end the day before we left to drive home to Tahlequah. The excitement of seeing our families made every mile go by quickly. Our plan was for both families to have dinner at Andy and Stephanie's when we arrived.

It was a wonderful homecoming. There were more hugs and kisses that one could count. We sat down to a feast. Stephanie and Grandma had prepared a great meal of fried chicken, homemade bread, coleslaw, mashed potatoes and pecan pie for our desert.

We were told we looked older and more mature. It was fun to talk as adults with our families. After dinner Maggie asked how the businesses were doing. It got quiet for some time before Andy told us that Rex had visited them a week or so before and they had a long serious talk.

Rex had told them that their employment contract would be ending on May 1st of 1953, which was less than six months away. His reasoning was that both businesses were doing so well under the current leadership, that neither he nor Grandpa would need to be tied to the businesses in the future.

This was a shock to us as we could not imagine that Andy and Grandpa would no longer be involved in the businesses. I questioned my role or that of Maggie's. Why would Rex need us if they were doing so well without our families?

Grandpa spoke up and said that he was getting bored with his role at work and at home. He said that all he did was wonder around and see that everything was working well. Then go home a tend a little garden and wait for dinner time. Andy expressed the same feelings. Grandma and Stephanie admitted that it was the same for them. After years of pouring themselves into our businesses they were suddenly no longer needed. Life was becoming boring.

Both families had money and comfortable homes, but neither seemed to have a purpose. Grandma spoke up and said that Rex had told them that he was working on a new venture and that he was not at liberty to talk about it now. He said his plan would include a place for them.

That made us feel much better. But for Maggie and me, it left our plans, for the future, uneasy. Why were we going to school when upon graduation we were to start running our family business if it was doing well without us?

For the first time in almost three years, I felt my future was uncertain. Being home with my family soon replaced the future with the present. It was a wonderful Christmas. Perhaps the best one in my life. Grandma and Grandpa were so happy and looked so good. They had no stress in their lives. You could see Grandma's touch throughout the house. It was her kingdom. Grandpa still had that old radio. Although it had been replaced with use by a modern version, it still occupied a place on an end table by his new favorite chair.

I believe they looked younger than ever before. The opportunity Rex had given them, changed their lives and gave them a way of life they could never have imagined. I was thankful for what he had done for all of us.

The day before we left to return for school, I had a chance to talk alone with Grandma. I asked her about Grandpa. She told me he was bored and needed a purpose. She said he sometimes just left in his truck and would drive around for hours looking at the countryside.

Then she told me that their lives had always been a struggle and they were always creating ways to meet their bills and survive. How they would save for months just so they could afford a new pair of boots or a new piece of machinery for the farm. She said, "Now we have the money for anything we want. We do not need to get up early or go to bed tired. But we need something to have as a goal or something that is interesting. We want to be near you and

Maggie. Seeing you a couple of times a year is not enough."

We said our goodbyes the next day and headed to back to school. Maggie and I talked the entire way about our families and the future. This had been the most difficult parting with our folks. Maggie said, "Gule, I have been doing a lot of thinking. I do not want a future that does not include living close to my family. I do not know what that will entail but that is what I want. I did not realize how much I miss them."

I told her that I felt the same way. We both agreed that we would need to speak with Rex about our future. The ride back to school seemed longer. We had an unsure future ahead and we were missing our families.

Shortly before the spring break, Rex contacted us and asked if we would come and visit Abby and him. Maggie replied for us and assured them we would be there the evening school let out.

The time to spring break went rapidly. The weekend calls home broke the school routine into bearable segments. I told Grandpa and Grandma about our upcoming trip to see Rex and talk about the future. As usual they were totally supportive. Grandpa never let on to me what Grandma had said about him being bored. Grandpa was never one to share with others his burdens. His faith was strong, and I guess he thought that between him and God they could work out most of the problems.

## CHAPTER TWENTY-NINE

Maggie and I left school as planned and made it to the ranch in record time. Abby met us at the door when we arrived at her home. We had the normal round of hugs and kisses before we even got inside. She said Rex was out on the ranch and would be home soon. We unloaded the truck and asked if we could clean up before dinner.

Abby said Rex was meeting the plane and would be home after that. Maggie was talking with Abby, so I went out on the front porch, sat in a chair, and observed the valley sprawling out as far as the eye could see. It was a beautiful piece of land dotted with cattle, horses, and the main ranch buildings.

My thoughts were interrupted by the approach of an airplane to the ranch airstrip. It landed in the distance, and I lost track of its movements. About twenty minutes later, I saw two vehicles winding their way toward the house. My first thought was that Rex was having guests for the evening. But why had Abby not mentioned that to us?

It would not take long for me to know the answer to my question. When the vehicles stopped, I could see several people inside. As the door opened and they starting getting out, I could not believe my eyes. It was Grandpa and Grandma in one truck and Andy and Stephanie in the other. Rex was looking at me with his Rex smile. I ran down the steps to hug each one. As we were greeting each other Maggie came out of the house and as it dawned on her who was here, she ran down to her mother and father. There were a few tears, and then some laughter. I worked my way over to Rex and hugged him. I said, "Well you really got me on this one." His only response was "Good."

After a few minutes Rex said that we should go in and see

Abby. I helped with the suitcases, and the girls went in to greet Abby. There were some more tears and hugs. This felt like a wonderful family reunion. For years to come, Maggie and I would talk about that special moment in our lives.

When everyone had settled into their rooms, we met in the great room and waited for the dinner to be ready. It was so wonderful to have us all together for this time. Rex told us that he would be sharing some ideas after dinner but to please Abby, we could not talk business at the table.

When Abby announced that dinner was ready, we all took a seat in the dining room. Rex asked Grandpa to pray before dinner. I do not think I will ever forget his words. "Dear Father In heaven, we thank you for your gifts to us. We thank you for all of us being together to share this time. Thank you for letting us feel what real love is like. Thank you for having a plan for our lives and thank you for Rex and Abby."

No one said a thing when Grandpa finished praying. It was just quiet for a moment. Rex finally said, "Well, I do not know about the rest of you, but I am hungry." That broke the silence and we shared a wonderful dinner and conversation. The food must have been great, but I hardly remember tasting anything. It was such a memorable dinner with all those who I loved in one room at one time and in the perfect place.

The girls helped Abby clean up and the boys sat in the great room talking small talk until we were joined by the girls. Rex took a long breath and started into a verbal journey of his vision for the future.

"It is wonderful to have all of you here with us. We have had a very successful experience with our companies over the last few years. They are still doing very well and that is happening without much input from any of us. There are good people in place, and I think that there are some opportunities coming up that could use our combined efforts."

"With that in mind I would like to share a little history and then some ideas. Oil exploring and drilling has been going on in Texas for more than 50 years. In this area, which is called the Permian Basin, the first big well was the Santa Rita No. 1. It was completed in May of 1923. It was drilled to a depth of a little over 3,000 feet. In the next 30 years, hundreds of well were drilled in this entire area. They ranged from 3,000 to as deep as 6,000 feet. "

"All the easy oil was discovered and pumped out of the ground. Production in these early wells started to decrease. Many of them reached production levels that were so low that it was no longer profitable to even run the pumps. This year marks the 30th anniversary of that first big well. In the last few months, a new approach to getting the hard-to-reach oil reserves, has been perfected. Many of these old wells are not running out of oil. They just cannot reach the oil that is trapped in some of the rock formations."

"This new method is called fracturing. It has not been tried in the Permian Basin. I believe that there is no reason that it will not work here. I know that some of this discussion seems strange to all of us, but we can learn and we can be part of this modern way of obtaining oil." He finally stopped talking and looked around the room.

No one said anything. The room was quiet. Finally, Grandpa spoke up. "Well Rex thanks for the history lesson. I am not sure how we fit into this future, but I for one would like to hear your ideas. I do not know about Andy, but I am bored not having something to do each day." Rex looked at Grandpa and said, "Roy do you have it in you for one more big adventure?" Grandpa simply said, "You bet I do." Rex smiled and said, "How about you Andy?" Andy said, "If Roy is in then so am I."

Maggie and I were sitting there wondering how all this would affect us. We were going to school to learn about running our family businesses. Now Rex was wanting to go into the oil business. It was all a bit confusing.

Rex was about to start into another explanation when Abby spoke up. "Rex, I think you have said enough for tonight. Let's enjoy are friends and you can take this up in the morning. It was hard to leave the subject at hand and go back to small talk the rest of the evening. Somehow, we were able to accomplish Abby's wishes. We went to bed with a lot of unanswered questions. As it turned out, most of them would be answered the next day.

We met for breakfast, and it was apparent that Rex wanted to get breakfast out of the way so that he could share more of his big plans. We did not spend much time on small talk, and we quickly consumed the wonderful variety of things Abby had prepared. When breakfast was over, we all gathered in the great room. Rex had prepared a folder for all six of us. He passed them out and then started speaking.

"As you can see these are figures and facts which were compiled from information available from the 1952 oil production numbers. There is every reason to believe that the costs will be accurate for the next couple of years."

He asked to turn to page three, there was a list of items he wanted to focus on during his talk. We were looking at the list as he continued to explain each category.
" These items are what will determine the profitability of oil production in the next few years."
Crude oil selling price is average of $2.68 per barrel. Drilling cost for a completed well is $80,000 to $100,000.

Recovering drilling cost would require the well to produce 30,000 to 40,000 barrels.

Oil field wages:
- Roughneck. $12-$18 per day
- Derrickman. $15-22 per day

Driller $25-$35 per day

Pumping price per barrel. $ .10- .25

Oil production lease, permits, and operating expenses $500 To $2,500 monthly.

Wells that were regarded as financially workable produced 1,000 barrels per day.

Purchasing a new rotary drilling rig $130,000- $150,000

4 ½" drill pipe $3.20 per foot.

Oil lease cost: Royalty lease 12 ½ %
- Non-producing/wild cat land. $1 - $5 per acre
- Promising but unproven $10-$50 per acre
- Proven areas $100- $500 per acre

When he finished with the massive lesson on the oil industry, we were all pretty much overwhelmed.

"In conclusion, with this new fracturing process, we could turn some of these existing wells into positive investments and we could drill new areas of our ranch that are considered very likely to have oil reserves beneath the surface. We could hire the wells drilled by a dependable company or we could purchase equipment, hire our own people and drill our own wells."

"The ranch has the money, but we would need to form a new company to invest in oil whether we are drilling ourselves or hiring it done. I want us to work together on this new revenue source." Rex paused and waited for our comments.

No one knew what to say. We had just had a lesson about 30 years of oil drilling, learned the cost of every aspect of the industry and were expected to ask questions.

Grandpa was the first to speak. "Rex what can we do for a company in Texas, when we live in Oklahoma?" Rex gave one of his famous grins and said, "I am glad you asked that question Roy. For us to be successful, we will all need to be here on the ranch. It will be necessary for us to do many things personally in the field. We will need daily meetings and access to each other."

"With that in mind, I propose that your families relocate here to the ranch." Is that something that you would be willing to do? Let me give you some time to think about that part. Here is what I propose. The new company could foot the cost of building a new home for each of your families.

They would be on the ranch and yours to live in as long as you want. I suppose that you need to make the decision on whether to move, before we go any further." Rex finished by saying, "How about we meet back here for dinner and you folks just enjoy the ranch and each other."

Rex told us that while we were talking, he had two trucks delivered to the house and they were ours for the remainder of our visit. Then he said, "I will see you at dinner." He kissed Abby and headed out the door.

I do not know what everyone else wanted to do, but I knew Maggie wanted to go see Precious. It turned out that everyone wanted to see Maggie's horse, so that is where we started our day of exploring.

After looking around for a while, we located Bobby Wilson, the head horse trainer. He was glad to see us and told Maggie where she could find Precious. Before we left, he said, "Maggie, I was talking to Adam a few days ago and he told me Precious spent two days in a corral next to where he was training some new cutting horses."

"She showed a lot of interest in what those horses were doing with the cattle. By the second day, she was trying to copy their movements. Adam thinks that her being a quarter horse and showing so much interest in what was happening in the other corral, is an indication that she may be a good candidate for cutting horse training. I told him that she was your horse and it would be your decision. Is that something you would like for Precious?"

Maggie did not need any time to think. She just said, "If she wants to herd cattle, then I am for it as well." Bobby said, "I think it would be good for both of you. I will tell Adam, and we will come up with a plan to expose her to what cutting life is all about. If it seems that she enjoys the experience, we will aim toward a new future for her. If she finishes her initial training, Precious will be worth more than a brand-new pickup truck. I have seen some well- trained cutters go for as much as $5,000."

We left Bobby and drove to see Precious. She was in a large corral with several other horses her age. When she saw Maggie, she came running over to the fence. Maggie crawled through the fence and gave her a big hug. Precious placed her head on Maggie's shoulder and held her tight. There was no question about those two having a special bond. I had often thought that Maggie was the only person Precious had ever seen with all that red hair.

Precious showed some interest in all the family members and eventually allowed all of us to touch her. She was two years old now and pretty much the size she would be for the rest of her life. Bobby had told us that Adam had already started her training for being ridden.

I took dozens of pictures of Precious with all the family members. Some of them hang on the walls of our home to this day. There were so many things happening to each of us in the new plan Rex had dreamed up. Other than our families moving to Texas, we were not even sure what our roles might be in the future.

After leaving Precious, we all drove around the ranch areas that were closest to the headquarters and airstrip. Andy said we should look for possible homesites if we were going to move to Texas. The rest of the day was not spent exploring. I must admit, it was fun spending the day with all the people I loved.

We took time out for lunch at the Chuck Wagon and then continued our search. There were dozens of places to build a home, but nothing came close to the area where Rex had his house. That hill overlooked the ranch for as far as one could see. It really was spectacular.

By the time we had returned to the house, Abby was already preparing dinner. The three girls pitched in and help her finish. The three guys sat in the great room waiting for Rex to come home. We shared our thoughts about making Texas our new home. Grandpa asked me if I was for the move. I told him that Maggie, and I had not talked about it alone. But I did say, "We have spent so much time on the ranch that it would not seem odd for us to settle here for the next phase of our lives."

Before we could go any farther with our conversation, Rex walked in the front door. He took off his hat, sat down by Andy and said, "Well how did it go today?" Andy told him we had gone to see Precious and spent the rest of the day exploring. Rex said, "Does that mean you are going to be moving to the ranch?" Andy laughed and told him that it was a very big step. He said, "A lot of our decision is based on what our kids decide. We do not want to be separated from them when they settle down after college. Which brings us to the point of asking what is your plan for Tommy and Maggie?"

Rex though for a minute before he responded. "Andy, I do not want to say much before I present the plan to the kids. But I will say it would make it necessary for them to live here." Andy told him that he and Stephanie would be much more inclined to favor a plan that included being near Maggie. Grandpa simply said, "I feel the same way that Andy does."

Rex was about to say something when Abby called out to us that dinner was ready. We washed up and gathered around the table. Rex would usually ask Grandpa to pray, but this evening he said the prayer himself. "Dear Father in Heaven, thank you for these friends who have become our family. Thank you for the gifts you have given us and the opportunities that lay before us. Help us always to be good stewards of what you have given us and to be fair and honest in all our dealings. Amen"

Dinner was wonderful. Maggie told Rex and Abby all the news about Precious. Grandma talked about how beautiful the ranch tour had been. Stephanie said that it had been one of her favorite times with our families. It was Abby not Rex who posed the question of whether we would be moving to Texas. It was totally silent after she spoke. Grandma was the first to speak. "Well, neither Roy nor I had ever given any thought to leaving our home. But the last three years of seeing Tommy for a few weeks here and there has not been my favorite time."

Andy spoke up next. " I do not know how Ida feels but Stephanie would like to get rid of me now and again. She is not used to having me around with nothing to do. I do miss Maggie every day of my life."

Rex looked at me and said, "How do you feel Tommy?" I answered him from my heart. "Rex I was caught off guard with this oil thing. I thought my future was running a business that my family started. My plan was to spend my time with them and with Maggie's family. If my future is to be on the ranch, it would have to include my parents and Maggie's."

Everyone was looking at Maggie. She cleared her throat and said, "I feel the exact way that Tommy feels. There is just one other thing. If we are going to be here, you will need to build a home for Roy and Ida, one for my mom and dad and one for me and Tommy. I have been waiting for him to propose to me, but it has not happened so I will just say that we will be getting married the summer that we graduate from school and we will need a home."

I guess that the news was not unexpected. But having Maggie deliver it this way was a total surprise. I guess it surprised me more than anyone. Grandma started clapping and said, "That is my future daughter-in-law talking and I totally support her!"

Rex spoke next, "That is the best news yet. It was what Abby and I hoped would happen. Not only will the ranch pay for your house, but we will also pay for your wedding."

Grandma looked at me and said, "Well Tommy, what do you have to say?" I looked at Maggie and said, "The day you can ride Precious, I will get Willie, and we will ride to a special place, and I will ask you to marry me." Maggie's eyes filled with tears, and she ran to me, and we hugged and kissed. I guess it was a pre-proposal.

Grandpa stood up and said, "Rex, I guess you have your answer. We will be moving to the ranch, and we do not even know what your plan is for our future." Rex was quiet for a moment. I think we all were emotional. He simply said, "Roy you are all going to love our plan. It is not mine any longer. It now belongs to all of us."

## CHAPTER THIRTY

The next morning, Rex explained what part of the plan he could see each of us doing. It was very well thought out and detailed. Once again, this man had shown us what an unusual person he was in every area. He was a master at seeing the strong points in other people. Many of his plans were unreachable or even unreal to the average person. For Rex, it was just his way of life.

While we all sat around the big dining room table, he rolled out maps and documents. Each was placed in some form of presentation, that only he understood. "I am thankful that you have all decided to make the ranch your new home. It not only will be much more convenient for our plan, but it is something that will make Abby and I very happy."

"With that in mind, one of the things we need to address is housing. Our cabins are nice but are too small and are for temporary stays. You have had a few hours to look around, and I am curious as to whether you have found a place you might like to have a home built." To our surprise, Maggie was the one who spoke up first. "Rex this entire ranch is beautiful. It would be very difficult to find a perfect spot that is still close to the headquarters and the airstrip. We have not had much time to look.

We have talked about among ourselves, and we believe you have the perfect location. We would never want to intrude on that privacy that you enjoy as you look out over the valley. We have decided to leave the location up to you and Abby. There is no lack of wonderful places where we can start the next phase of our lives."

In many ways we were all caught off guard by Maggie's comment. We had not come to a group opinion on the subject. The way Maggie expressed her words was more than perfect. You could see that her comments had struck a soft place in every listener's heart.

Abby spoke next. "Thank you Maggie for those wonderful words. Rex and I feel very blessed to wake up each day and be able to sit on our porch and look out over all that God created and loaned to us for a season. I do not know how Rex feels about it, but I would love to share this experience with all of you. Our hill is almost 2 miles long and there is more than enough room for all of us. Nothing would bring me more joy than to look out my window and see the people that I have come to love and have them enjoying what we see every day."

Rex was the next one to speak. "Abby is right. This is the natural place for you to be. We can build the houses a quarter of a mile apart. That gives each family some privacy and still allows us all to be close to each other. I think it will perfect."

I guess the decision was made. No one else spoke until Rex said, "So that brings up step number one of our plans. I think that all the girls should get together and design the three homes. I have a contractor who owes me a favor. I believe we can count on him to make this a priority. We can get an architect friend of mine to draw up the plans and help us with site preparation. I want Maggie and Tommy to finish their degrees next year. I will have Andrea do some work with A&M and see if we can replace some electives with geology and some other classes that would be specific to our new plan."

"This summer Maggie can help with the house designs. As far as Andy, Roy and Tommy, we will need to sit down and work out some specific roles for each of you. When the kids return to school, you can use that time to return home and make plans for what you will do with your homes in Oklahoma. I think we should plan on all meeting here again when the school year ends in a few weeks."

Everyone agreed and we were all excited. It was going to be a wonderful adventure, and our families would not be separated for any time in the future. The long periods of time that I had spent away from Grandma and Grandpa had been very hard on me. I really did not care what Rex had planned for me. My future would include spending the rest of their lives with them in this beautiful place.

There were only a couple of days left before we would be leaving for school. We spent that time with our families exploring the hill for homesites, sharing time together with them and enjoying every evening with Rex and Abby.

Maggie and I had very little time alone. One afternoon we were able to slip away and go see Precious. On the way there, I asked her about her announcement about us getting married in a little over one year. She just laughed and said, "If I had to wait on you, I would be old and ugly before getting married." I told her that she was the love of my life and she would never be old and ugly. I told her that I was sorry for not having shared with her how much I loved everything about her.

Maggie told me that she was used to me keeping my feelings inside. But she also said that when we had children, I would need to share my feelings with them. Maggie was my guide in life. She always seemed to know what to say and how to say it so perfectly. I could never picture my life without her. It was beyond my dreams that in a year we were to get married, have our own home, and start a life completely different than what we had planned for in the last three years.

On afternoon, I was able to get Grandpa alone so we

could talk. I thanked him for being my grandpa and loving me for my entire life. I told him how much it meant to me that we would be living and working together for the rest of our lives. As usual, he thought for a long time before he answered me. "Gule, it is I who am thankful that the Creator sent you to us. You will never know what a gift it has been to have you in our home and in our lives. To have this chance to be with you and Maggie as you start this next part of your lives is more than we could have every asked. I will die a happy man. The Creator never makes mistakes. His plan is perfect, and Grandma and I are blessed to be part of it."

I will not forget those few hours with him. I had never met a kinder person in my life. He had lived through struggle, poverty, and now comfort. But none of those ever changed who he was and the man who I loved.

The day came when Maggie and I had to leave for school. It was bitter-sweet. We had been able to enjoy several wonderful days with those we loved. But we had to leave them once again. This time, unlike the others, was for a short time and soon we would all be together again working for a new life in a new place.

Rex and Grandma took lots of pictures and we had our customary hugs and kisses. As we drove down the hill toward the headquarters, Maggie said, "How can we be so blessed to have such loving families?" She had some tears in her eyes. I could not answer her because I was choked up and did not want her to see it. The trip was nice because we were alone. We did not need words. Just being together was enough.

The rest of the year was busy. We had term papers due and then the finals in every class. Both of us still called home every weekend. Our friends from the ranch, Darren and

Jane, were both graduating that spring. Darren was moving to Dallas where he would be working at an accounting firm. Jane and her fiancé Kevin were moving to San Antonio where they would be working for an earth moving company owned by his family.

It was hard to believe that three years had gone by so quickly. At the same time, there were so many major changes to our lives. In one more year Maggie and I would graduate from college. According to Maggie we were getting married and starting a new life in our new house on the ranch. I believe it would have been difficult to fit more changes into those three years.

Our families had their own adventures which started the day they met Rex Walker. Looking back, I think each of our parents had grown in ways that they would never have had the opportunity to do, if their lives had centered only around our little town of Afton. The best part was that this next phase of our lives would be us all working and living together in a wonderland called Walker Ranch.

Rex sent our pilots and the plane to fly Darren's and Jane's family up to the school for their graduation. He never seemed to forget his "people", as he called them. I would never learn how his life had been shaped and how he came to be the brilliant and loving man who we knew as Rex.

# CHAPTER THIRTY-ONE

Maggie and I had already packed the truck so we could leave as soon as the graduation ceremony was finished. We had said our goodbyes to Darren and Jane and drove back to the ranch. Abby had called Maggie and told her that our families would not be back to the ranch for a week or so and that she and Rex wanted us to stay with them.

We arrived at the ranch in time for a very late dinner. This was the first time we had shared an evening alone with the Walkers for a long time. It was like coming home to family. We did not talk about business. Enjoying our time together was wonderful. Rex did tell Maggie that Adam wanted her to come and see him in the morning. Rex could see concern on her face, so he assured her that there was no reason for her to be alarmed.

Rex told me that while Maggie was going to meet with Adam, he wanted me to spend the time with him. It had been a long day, so we said goodnight and agreed to meet up for breakfast. Rex said that Abby and Maggie could fend for themselves and that we would get an early start and have breakfast at the Chuck Wagon.

By 7 a.m. Rex and I were already having breakfast. I had not been alone with him for many months. If my memory served me correctly, every time we were alone, he used it as a teaching session. Today would be no exception. I learned that we were going to some property near San Angelo and that our day would be spent learning about oil fields and drilling procedures. I was exciting to me to be exposed to something unrelated to anything in my life thus far.

On the drive to our destination, Rex asked me if Grandpa and Andy were really interested in his ideas or were just going along with the plan to be near Maggie and me. I assured him that each of us had our own reasons for being interested in the new life that was being offered to us. It was going to be wonderful having us all together in one place, but that alone was not enough to create the excitement I had seen in our families.

That seemed to satisfy his question. I think he just wanted reassurance that he was not pushing us into something that he alone was excited about for the future. I did my best to reassure him that Andy and Grandpa were excited about doing something new and waking up each day with a purpose.

He paused for a few minutes and then asked, "What about you and Maggie? You have given up three years of your lives studying subjects which would allow you to take your families companies to new levels in research and sales. Now I am changing the whole plan at the last minute and asking you to learn about a completely new industry. Are you and Maggie really on board with my plan?"

I told him that we did not even know his whole plan, but that he could rest assured that we were in support with what he had told us so far. I told him that we had total confidence in his choosing the path for our future and that would never change.

I said, "Rex, Grandpa told me long ago that you could measure a man by his actions. You are a man who cares more about people than profits. One of the goals of anything you seem to get mixed up in, is what good will it do for your people. You enjoy trying to make everything better. The satisfaction you derive from your projects does not seem to be measured by profits. I do not know what drives you to constantly be involved in all your endeavors, but I can tell you that it inspires me to be more than I ever imagined that I could be. Thank you."

Rex suddenly found something of interest outside his window as we were driving along. I could not see his face for some minutes. When he finally spoke, it was short but there was emotion behind the words. "Thank you, Tommy. You have no idea what your words mean to me."

He was quiet for a long time. We just drove along and enjoyed the scenery. There was ranch land as far as you could see and as we got closer to San Angelo, we started to see more oil wells and drilling rigs. Rex finally spoke as he pointed out the window. "That is why we are here today. Our future is going to be in the oil industry. I hope that long after I am gone you and your children will build on what we are going to accomplish in the next twenty years."

Once again, this amazing man was planning people's lives before they even realized it themselves. I had brought along a tablet, a couple of pencils and my camera. This was all new to me and there was zero possibility that I would be able to keep up with all that Rex would be saying without taking some notes.

By early afternoon, we had arrived at a drill site where it was apparent that Rex knew some of the crew. It was all strange to me. The huge drilling tower was surrounded by a drilling platform, piles of drilling pipe, storage tanks, all kinds of support equipment, and what looked like a mobile work shed.

Rex led the way and we walked over to the drilling platform stairs where we were greeted by one of the crew. He and Rex shook hands and Rex said "Eddie it is great to see you. I would like you to meet Tommy Meadors. Eddie Most had hands calloused from years of hard work. As he shook mine, he must have wondered if I had ever worked at any job. But he smiled and said, "Welcome to my office."

Rex said, "Eddie as we talked about on the phone, I am considering drilling on some of my ranch. We stopped by today so we can learn a little about oil drilling rigs. Tommy will be helping me plan the future." Eddie smiled again and said, "Well Rex let's have a tour." He walked over to a tool locker and returned with a hard hat for each of us. "Safety first."

We spent more than two hours with Eddie showing all the different operations of the rig and talking about the different roles each member of the crew must fill. I had several pages of notes by the time the tour was over. Before we left the site, Rex and Eddie went over by one of the storage tanks and had a long conversation.

When they came back to where I was standing, Eddie stuck out his hand and said, "It was nice to meet you, Tommy. Maybe we will see more of each other down the road." I shook his hand and said that I would look forward to that happening.

The entire trip home was dominated by talk about oil wells and opportunities. During one slow time in our conversation Rex asked what I thought of Eddie. I told him that it seemed to me that he knew everything about drilling for oil. Rex agreed. "When we move forward with our plans, I want Eddie to be part of our team."

He then told me the story of Eddie Most. He was a Cherokee Indian from Oklahoma. He had worked his way to a top job while enduring discrimination toward Indian people. He had earned respect from other workers by being fair and rewarding the loyalty of others. Eddie never talked about his early life. He was a very private person.

Rex asked me the oddest question. "Tommy, do you think that Grandpa being Cherokee, could help him become a friend to Eddie?"

I did not understand where he was going with his question. I told him that being Cherokee was a good reason that they might get along. Rex saw something in Eddie and Grandpa that I was unable to recognize. As usual, he was looking at Eddie as more than a future employee. He wanted to know the person, not just the worker. He was going to be looking to Grandpa to break down those walls and allow Eddie to be proud of who he was and where he had come from.

We arrived home in time to clean up and get ready for dinner. Maggie wanted to know what we had been doing all day. I told her about the trip to the drilling site and meeting Eddie Most. I wanted to know what she found out from Adam. It was easy to see that she had been waiting to tell Rex and Me all about her day.

"Adam and I met at the training corrals. He told me he had a surprise for me. We walked to another pen, and I could see Precious, standing in the pen and looking in our direction. To my surprise she had a bright blue halter on her."

"As we got closer, she began to weenie and pace back and forth by the fence. Adam and I crawled through the fence and Precious greeted me with affection. Adam was carrying a s short rope which he attached to her halter. Then he led her around the corral, had her back up, and start and stop on command."

"Adam handed me the rope and told me to work with her for a while and then we would do some grooming." Maggie was beyond excited at the changes in Precious. She went on to tell us how well behaved Precious had been while they groomed her. Adam had told her that in the next couple of weeks they would be starting the light training for her first experience with having a saddle blanket on her back. That would be followed with having a saddle placed on the saddle blanket. Then she would be introduced to a bridle and a bit.

He felt that by the time she was 30 months old, he would be able to do some very light riding on her. Adam told Maggie that she would be able to ride Precious for short periods of time within the next couple of months. The last thing she told us was that Adam was going to put Precious in a pen with some cows the following morning and that she was going to go and watch.

I do not think Maggie took a single bite of the dinner during the time that he was talking about Precious. Rex and Abby just smiled and listened. Neither of them could have possibly known what it meant for them to give Maggie that little horse.

While the girls cleaned up from dinner, Rex and I sat on the front porch and talked about our day. He was interested to know how I felt about the change of plans for my future. I assured him that I was excited about learning about the oil business. I felt that it would be more interesting than working in the meat business. He told me that he was looking forward to learning all he could about oil drilling and marketing.

After thinking for a couple of minutes he said, "Tommy, each of us reach a stage in life where we question if we should just sit back and enjoy what we have or what we have accomplished. I am not made that way. I need a challenge. I need to be surrounded by people who are willing to step out and explore the next adventure with me. I need a reason to get up in the morning."

"I love my ranch, the folks who work with me, and the other businesses that we have helped grow and prosper. But I need a challenge. I have been watching this oil thing for many years. It has been interesting to see the successes recovering oil have my interest." and, also the failures of oil pioneers.

"For the first time in the history of the oil business, the risks can be mitigated to the point where there is more success than failure. The mistakes of those who came before us have helped shape the future in a good way."

"I have told Abby that this is my last big adventure. She does not believe me, but I am serious. I could never have had the life we experience without her support. She deserves to have more of me than she normally gets.

"I never realized how buying your family businesses would change my life. All of you are the perfect fit in our lives. There has always been something missing. You all have filled that empty place. Abby has never been so happy, and I have never been this content."

"There will be a role for each of us in this plan. It will be exciting to have us all work together to create our future in this crazy world of oil."

When he finished speaking, I told him how Grandpa had been so excited about a new project. Andy was also ready for a new challenge. He was younger than Grandpa and was not ready to just sit back and watch life roll along. It seemed to reassure Rex that we were all excited to be part of his dream of doing well in the oil business.

The girls finally came out to join us, and we all shared a beautiful Texas evening on the front porch. Rex and Abby were cuddled up on one couch and Maggie, and I were cuddled up on another. It was a wonderful evening spent with a couple of people we had learned to love. The night sky blessed us with a display of stars that was almost magical. Life was good.

## CHAPTER THIRTY-TWO

The next morning, Maggie left to go see Precious and Rex and I grabbed a roll of maps and drove off to the part of the ranch which had every indication of being an unexplored region of the Permian basin.

Abby spent the day with Maggie and Precious. Those two were becoming very close. It was a relationship that Rex encouraged. I was getting a crash course in oil field history. After two days with Rex and the one day with Eddie, my head was running over with facts and information. There was no question that it was very interesting. I had never seen Rex so excited about anything in the years we had known each other. It was great.

That evening after dinner, Maggie and I were sitting on the front porch talking about how oil wells and horse training were so far from the things we knew as children. And now they were different from what we had gone to college to learn. We were so blessed to have these opportunities. The common denominator was our friends and benefactors Rex and Abby.

For the next week, I spent every day with Rex and Maggie spent every day with Abby. We met for dinner and that was about it. It was a good busy. My learning curve was increasing, and Rex seemed pleased with the progress. Abby was working with Adam and she and Abby went into town a couple of times after the morning training with Precious.

Maggie and I had each called home and talked with our families. Grandpa told me that he and Andy were working on some plans for their properties. He also mentioned that they would probably not be coming to
the ranch for a few more weeks. I was disappointed but

knowing that we were going to all be together in the future made waiting to see them a little easier.

One evening at dinner Rex announced that he and Abby were flying to Dallas for a few days. They had some business to take care of and then were making a fun trip out of it with some personal time together. Rex told us to enjoy our time alone. They left the next morning.

Maggie asked me what we should do that day. I told her we needed to go exploring and find a special place where we could visit when Precious was ready to be ridden. Her face lit up because she understood that it would be the place where I was going to propose. We went to the Chuck Wagon for a late breakfast and while we were there Maggie packed a lunch for our little trip.

Our exploring trip became a little longer than we had anticipate. On day four of our search, we saw a line of cottonwood trees in the distance. As we got closer, we could see a small lake which was fed by a creek. On the near side of the lake was a huge cottonwood tree. It was so much like our lake at home that I knew this was the place. As we approached the shore Maggie looked at me. There was a tear on her check. She grabbed my hand and said, "This is the place Gule, we do not need to look anymore."

We sat on the bank of the little lake and held each other. No words were needed. This was our new special place, and we were going to make this ranch our forever home. Maggie and I made a trip to this place many times, over the years. It never failed to remind us of our beginning here in Texas.

Rex and Abby returned the next day. Maggie had to tell Abby all about finding our special place. Those two were more like sisters than "parent" and child. I was happy for Maggie. The days were flying by, and we needed to start thinking about our last year of college, which would be starting soon.

I spent the next day with Rex. He returned from his trip with a more focused approach to the new plan. His ideas were not hurried. The plan was carefully thought out. It must have given him some encouragement as one by one the pieces were coming together in his mind. I was overwhelmed at times because he shared so much with me. It would take me many years of working with Rex to even begin to understand this man and his mind.

We spent the last few weeks of summer, planning the next steps for our future business. Rex had very specific roles for Grandpa and Andy. He wanted the girls to concentrate on getting the housing started. Each day we made a little progress in the planning stages. He refined his plan for Grandpa which was for him to develop a relationship with Eddie Most and convince him to come to work for our company. He wanted Andy to study the equipment inventory needs if we decided to have our own drilling company.

Rex was wanted more facts and figures before making the decision of where to drill and who should drill. Andy was a good fit after running his lumber mill and our companies inventory needs. Grandpa brought his own slow and steady way of getting things done without hurrying and having to redo any mistakes.

We were basically combining the wisdom of three generations. Each of us was allowed to express our opinion and it was always well received by the others. Rex wanted us to come up with a new company name because there was a mountain of paperwork that his lawyers would be working on in the coming months. They needed a name for those documents.

Rex made a call and asked for our two families to come to

the ranch for some meetings before Maggie and I needed to leave for school. They arrived in a couple of days. We all met after they had settled in on the first evening. Rex was chomping at the bit to get things moving along. He told the girls that he had scheduled a meeting with the contractor for the following morning. He wanted them to approve of the final site for each of the homes. Then he asked them to start giving the builder the floor plans of the homes so he could have an architect start drawing up some plans.

It was a big project for them, but Rex wanted it done exactly as they wanted it done. So, step one was assigned to the girls. Step two was a name for the company. Rex asked each of us for a suggestion. In the end we decided on WSM Oil Company. It was short for Walker, Scott and Meadors oil company. Step two completed.

Grandpa was to meet with Eddie Most the following afternoon and start working on him to become a member of our new company. Andy was to visit the site with Grandpa to take a complete inventory of all the equipment, and inventory at the drill site.

I was to spend the day with Rex, and we talk about the plan for me when I graduated. Rex did not want me to confine my learning to any one area of expertise. His plan for me was to learn everything about the oil industry from exploration of undrilled land, to selling our oil when we found some reserves.

He wanted me to understand every aspect of drilling, pumping and storing oil. It was a tall order for a twenty-year-old young man. That was his plan and he wanted me to be ready to take over the company at some point in the future. In one way, it was exciting. In another way, it was more than I was able to process so quickly.

Rex never hesitated with his plan moving forward. In his

mind, each of us was part of a great adventure. He needed all of us working together to reach his goal. In the next few weeks, we were all busy. The site prep for the homes was completed and roads were being built. Utilities were planned for servicing each house. The girls had worked with the architect, and floor plans had been developed.

Grandpa and Andy spent many days at the drilling site or recording facts and figures regarding drilling. I felt that Rex would be using all the info to decide sooner than later, to hire drilling or form our own drilling crew and do it ourselves. Grandpa still did not have a positive answer from Eddie. But he felt that he was getting closer to that goal.

The day arrived for Maggie and I to leave for our last year of school. Andrea had done her magic, and we both had some classes that would be helpful in our new roles at the ranch. We left with a different feeling than we had in the other years. In the spring, we would be coming home to our families. We would all be graduating into something new together. It was exciting.

The first semester passed rapidly, and we headed home for Christmas. Maggie was so excited she wanted to see her parents, the progress on our new home, and of course Precious. On the trip home, she reminded me that she would be able to ride Precious to our special place and that I needed to propose so that we could start planning the wedding for the coming summer.

I told her that I was thinking the same thing. She seemed happy with my answer. I was a little scared. A new job. A new wife. A new home. It would be a complete change to every part of my life. Maggie did not appear to have any of those thoughts. What a wonderful girl. She was going to be my partner for the rest of my life. I was very blessed from the time she stood up for me in elementary school to now when she was about to become my wife. I could not imagine life without her.

Christmas 1953 was the best holiday of my life. Rex had a huge Christmas tree in the great room. Grandma had helped Abby and Stephanie wrap presents. Grandpa and Andy helped Rex with last minute plans for flying home and giving our employees their Christmas bonus. Maggie and I planned our little trip to our secret spot that was no longer very secret.

By Christmas Eve, the guys were back from Oklahoma, and we all spent a wonderful evening opening presents and sharing the moment with each other. I had contacted Adam and asked him to have Precious and Willie ready for a late morning ride. Maggie had packed us a lunch from the Christmas Eve leftovers.

We met Adam at the training arena. This was going to be the first time Maggie had ever been on Precious. Adam helped her mount up after she had loved on Precious for a few minutes. It was like they were both born for this moment. Precious was so patient and allowed Maggie to take all the time she needed to get comfortable. Willie was just Willie. He was always ready for anything. Maggie thanked Adam for all his hard work and training of Precious. She had tears in her eyes as she leaned forward in the saddle and hugged Precious. I think it got to Adam as well. He just kept looking at the ground and scratching a line in the dirt with his boot.

When all was ready, we slowly walked our horses toward our secret place which was a few miles away. Adam had placed a pair of saddle bags on Willie and there was plenty of room for our lunch. I was impressed with how gentle Precious was with Maggie. She responded to any movement of the reins.

Maggie was in heaven. Her little baby from a few years ago was maturing into a wonderful companion. Adam had told us that she had been in the pen with cattle and other horses who were being trained. She still showed the same keen interest in cutting. He had even allowed her to pick her own way through the steers. He continued to give her gentle directions but allowed her to try doing it her way.

He had told Aggie that each morning, Precious was standing at the gate to her stall, obviously wanting out to go to work with the cattle. She needed a lot more training, but she was learning quickly and seemed to really enjoy the experience.

We arrived at our place a little afternoon. It was a very nice day considering it was the end of December. Maggie had brought two blankets for us to sit on during lunch. I got them out of the saddle bags and Maggie got the lunch. We tied the horses to a tree that was surrounded by some nice grass. We decided to let them eat for a few minutes before taking them down to the water.

Maggie wanted to rest for a bit before we ate lunch. I think this was my clue to start the expected proposal. I did not want to disappoint her, so I started with what I had memorized. "Maggie Scott, you were my first friend when I started school. You became my best friend by believing in me. You encouraged me and tried to protect me."

" I do not know why you cared about a little farm boy from a poor family. You were the reason that school was important to me. I knew that each day, I would see my little friend with the flaming red hair.

"As the years rolled along, I have so many memories. The first time that we held hands. The first time you kissed my cheek. The first time that I watched you cry. The first time

I felt something inside when I had not seen you for a while. The first time you kissed my lips. You have always been my princess, and I cannot imagine a life without you by my side.

"Will you marry me and be my wife?" There were tears in her eyes as she whispered, "Yes I will Tommy Meadors". That sealed the deal. All that was left was to give her the ring that Abby had helped me find. I dug into my pocket and got the little box. She opened it and I helped her put on the ring. She just held out her fingers and looked at that ring for a few moments before she said, "This is the happiest day of my life." We kissed and held each other for a few minutes. Maggie remembered that we had not watered the horses. We untied them from the tree and walked them down to the lake. While they were drinking, we just held hands and enjoyed the moment.

The silence was broken when Maggie said, "Gule, why did you not go on the play baseball in college? You were a great player." I thought for a moment before I answered. "Maggie, to me baseball was something that Grandpa and I shared. Those hours and days were a special time when all of life's burdens were shut out, and it was just the two of us reliving a part of his life. We shared something unspoken as we threw the ball or practiced batting. Nothing could ever equal that feeling. It was not important to me. Grandpa made it special. I hope he lives long enough to create that same feeling in one of our boys."

Maggie laughed and said, "One of our boys! How many are you planning to have?" I laughed and said, "Enough to have our own baseball team." We finished with the horses and decided to have lunch before we headed back to the house. The food could have been terrible, and it would still have tasted like a feast. This was a perfect day with my red headed beauty.

We packed up and slowly rode back to the house. Abby met us on the front porch. She looked at Maggie with a question on her face. Maggie just nodded and Abby ran and grabbed her and held her for the longest time while she kissed her all over her face. A person would have thought that it was her own daughter. In a way, I think Abby felt that way about Maggie.

The rest of the day was just as special as the first part. Everyone was so happy. It was like having Christmas a second time in the same week.

After dinner, the girls all stayed in the kitchen and were making wedding plans and finalizing things about our new homes. The guys were in the great room talking about oil. Rex was happy with how things were coming together. He still wanted to move things along faster. He was a man on a mission, and he would not be satisfied until he saw a well of our own, pumping oil and gas.

The day arrived when Maggie and I left for our last semester of college. This was the first time there had not been tears. Just smiles and good wishes. It was the end of 1953 and the beginning of 1954. We would be graduating and getting married before the calendar showed 1955.

## CHAPTER THIRTY-THREE

The trip back to school was different this time. We were getting closer to a new phase of our lives. We had learned so much. We matured during our time at A & M. Very soon, we would be assuming the role of adults with a future in business. Childhood was being left behind. Maggie and I were surrounded by the best support group one could imagine. I guess we could say that we were living a real-life fairy tale.

After we had been driving for a few hours, Maggie started talking about the wedding. "Gule, they all want us to have the perfect wedding and then the perfect honeymoon in some exotic place. I feel terrible inside because those are not the things that I want. Will you be upset with me if I told you what I would like for our wedding?"

"Maggie, it is our wedding and you are the special person. You are the bride. I will support anything you want for your special day. You know me so well. I am a simple person who is happy with simple things. Tell me what you want for your wedding."

Maggie looked relieved. She kissed me on the cheek before she started to talk. "Gule, I love you and I do not need some big fancy wedding to make me happy. I was thinking that we could have a simple ceremony in the great room at Rex and Abby's. To be in that room with all the people I love would be my best wedding gift. I cannot think of a better place to have a honeymoon than in our new home which should be completed by the wedding date. "

When she finished talking, she looked at me to see my reaction. I just smiled and said, "As usual, you have made the perfect choice. I am proud of you for expressing how you feel. Abby and your mom just want to make that day as special as possible. When you explain your feelings to them, they will be happy to help carry out your wishes."

"Thank you Gule." It was settled. All that was left to do now was tell our families about our wishes. It was all going to work out just fine. Deep inside, I knew that Rex would have to do something special to leave his mark on that perfect day. I just had no idea what it might be.

We got settled in at our dorms and decided to go out for some late dinner. It ended up being a hamburger near the campus. Maggie was tired and happy. A few short months and it would all come together.

We still called home every weekend. Maggie got more phone calls because the girls were still working on house related decisions. Our weekends were still busy and there was not a lot of alone time. It was ok because we knew as each week went by, we were closer to our new lives.

Rex wanted us to come home for spring break. There were a lot of things that were pending. It was easier to deal with them in person. That is what we did. It was dark when we arrived at the house. Grandpa and Grandma were staying at one of the cabins. Andy and Stephanie had gone to Oklahoma to take care of some business. We had a nice evening with Rex and Abby.

The next morning, we met Grandpa and Grandma at the Chuck Wagon for breakfast. Maggie told Grandma what she wanted for her wedding. Grandma smiled and said, "Maggie you are more special than you could possibly know. We are going to make your simple wedding into something special. Every day I learn a little more of why Gule loves you so much."

When we finished, Maggie and Grandma left to meet Abby at the house. Grandpa and I were to meet Rex at the ranch headquarters. As we left for our respective meetings,

I looked at the two people who had loved me from the time I was a baby. They looked younger and more vibrant than I could remember seeing them in recent years. It made me so happy to see that they had a purpose and were adjusting to the life on this ranch.

Grandpa and I got to the office before Rex, so we just chatted until he arrived. He was carrying a handful of papers. Rex did not waste any time before asking Grandpa how things were going with Eddie Most. "Well Rex it is not going as fast as you would like. Eddie is happy where he is and to make a change, he would need some interesting changes in his life. I have been going slowly and trying to gain his trust. At our last meeting he told me that he would give me a list of his demands and explain each of them to me. I was hoping to get the list next week when we see each other."

Rex asked Grandpa if he knew any of the things Eddie wanted. Grandpa said that he did not. It was easy to see that this was frustrating for Rex. He was not the type of man to sit patiently by while someone made up their mind over some issue that was not important to Rex.

Grandpa told Rex that he needed to be patient. "Indians have a different speed than white folks. What may need immediate attention by you, may not be a crisis to them. It is better to go slow and win the race than to hurry and mess it all up." Rex appreciated Grandpa's counsel even if it was contrary to his own nature.

The next item was a report that Grandpa and Andy had been completing. It was a very detailed list of items needed for a drilling platform and support items. Grandpa handed it to Rex and for the next couple of hours we went over each line on the report.

When we had a break, Rex asked me how the wedding

plans were going. I guess this was as good as any time to tell him what Maggie wanted. When I finished, he took a moment and said, "Well whatever you guys want is good enough for me, but it is going to be hard for Abby to hold back. Maggie is like her little sister that she wants to shower with everything special. I will do my best to keep her under control." We all laughed.

I told him that Maggie would tell Abby what she wanted. Convinced that we three were off the hook, we went on to finish some other business. Rex told us that all three families would be attending our graduation. He planned on flying everyone down for the ceremony.

Rex finished the meeting saying how excited he was that we were all taking this next adventure together. "We are going to have a great team made of our families. It is going to be one heck of a ride."

As we left, Rex said he would see us at dinner. Grandpa asked me if I had anything pressing. I told him that I did not. He suggested that we go over and see how things were going on the house construction.

It was just a way for us to be alone. I think we both missed our baseball times together. We spent the next couple hours just being together. The progress on the houses was impressive. So much had been done in so little time. There was no doubt that they would be done by early summer.

I asked Grandpa if he was ok with the things Rex was having him do. "Gule, I am having more fun than I have had in years. When I visit Eddie Most, we often talk to each other in Cherokee. That way our conversation remains private. It makes me feel young again. Eddie brings out the Cherokee part of me. I find myself thinking in the Indian way. It has been a long time since I felt comfortable with that part of my life."

I told him that it made me so happy that Grandma and he were going to be living next door to Maggie and me. I told him that I wanted our children to have time to know and love my grandpa and grandma. I told him that it would mean more to me than he could ever imagine, if he would play baseball with my sons. He did his grandpa thing and would not answer right away. He even had to look away for a bit. He was having an emotional moment. I reached over and hugged him and told him how much that I loved him.

He finally said, "Grandma and I thought we might lose you after you graduated. The opportunity Rex has given us is a blessing from God." I put my arm around him, and we walked back to the truck.

I looked at my watch and realized we still had a little time before dinner. I asked him if he would like to go see Precious. That is what we did. Adam was riding her when we got to the training area. He was gently walking her through some exercises where she was cutting a steer out from the herd. He smiled when he saw us and kept up the lesson for another half hour or so. When they were finished, he rode her over to us and we patted her and gave her some attention.

Adam told us she was doing well and each week he was adding some more training. He felt that there was no reason that she could not become a competitive cutting horse within a year.

That evening after dinner, Maggie chose to tell Abby about her desire to have a very small ceremony. Abby's only response was that it was going to be the best small ceremony ever. We had a big laugh over her comment.

In the time we had remaining at the ranch, the girls decided on the exact date of the wedding. It was going to be the first of June 1955. The men made a plan that required some additional research and analysis. Rex was way passed the rest of us in his action ideas. He had an additional list of things to accomplish before Maggie, and I graduated and returned to the ranch.

Andy and Grandpa were fully engaged in everything Rex had on his timeline. They were going to be busy. This big plan that Rex had devised was giving each of them new life. It was exactly as Rex had described it, a great adventure.

The day for our trip back to school had arrived. We had a great sendoff. Our families had breakfast together and then lots of hugs and kisses. It was only a few weeks until graduation. They were going to be very busy. Maggie and I were very excited. It was almost over. Four long years of study and separation from our families. We had grown so much from our lives in our little Oklahoma town. I did not want our future to change who we were. I had no idea what was in store for each of us.

## CHAPTER THIRTY-FOUR

So much happened in the next few weeks. It was all a blur. Maggie and I completed all our exams , and our grades were very good. We had participated in a rehearsal for the graduation ceremony. Everything from our college days was packed in our truck. We were ready to leave as soon as the graduation was over.

The big day came. We met the plane when it landed. This would be one of the most significant days that our families had every experienced. Rex had transportation planned for our families. We drove back to the campus together and had a short visit before Maggie and I had to go to our assigned places. Rex and our families worked their way to the assigned seating.

It was a large group. Of the 6,837 students enrolled that year, there were almost 1,500 graduating. They called your name and you walked across the stage and received your diploma. The entire ceremony was a couple of hours long.

When it was over, Rex had a photographer take lots of pictures with our families. He thought of everything. Maggie and I just wanted to head toward home. We change our clothes and then followed them to the airport. As the plane lifted off, we headed to the highway and took what we thought would be our last trip toward the ranch from A & M.

We had been driving for a while when Maggie kissed me on the cheek and said "Gule, we did it!" I smiled. Yes, we had done it. Now the next thing before us was our wedding. I had not really kept up on all the details that the girls had been planning. But now it was time to get serious about this next step.

When we arrived at the ranch, we could see three new

homes in the distance. They were beautiful. We would learn from our families that there were still some interior things which needed to be completed before we could move into them.

Maggie wanted to go see our new house before we even went into the ranch house. Rex said that he would prefer that we waited for a few days so it could all be completed before we saw the final build.

There was still a little over two weeks until the wedding. We were hoping it would be ready for our honeymoon. Rex told us not to worry that he had been promised that all three would be ready by that time.

We had a wonderful evening with the people we loved. Rex and Abby were as proud of us as our own families were. I wondered how it was that we were so blessed. At one point in the evening, Rex proposed a toast to Maggie and me. Since none of us drank alcoholic beverages, Abby served sweet tea. It was such a wonderful evening. Maggie and I were very tired, so we excused ourselves and went to our rooms. As we kissed goodnight, Maggie said "Can you believe all this is happening?" I told her it was like watching a movie. It did not quite seem real.

The next day the girls worked on the wedding and the men had breakfast at the Chuck Wagon and then had a long meeting at the ranch headquarters. Grandpa had a hand full of papers, and I could see that Rex was anxious to hear what information they contained.

Rex said, "Well Roy, I hope you have some good news for us." Grandpa started his usual way, "I guess that will be up to you after you hear my report. As we all know, I have been working on Eddie Most and trying to convince him that he needs to be part of our new company. Finally last Friday, he handed me a list of things he would want to see addressed if he is to leave his job and come to work for us."

Grandpa proceeded to explain each item on the list and why each one was important to Eddie. The first item was that Eddie wanted profit sharing for him and the members of his crew. No one objected to that. The second item was that he wanted to be responsible for hiring the crew members. No one objected to that.

Item three was that he wanted to hire Cherokee people. He told Grandpa that during his entire oil field experience, he had seen the Indian people discriminated against. They had to work harder than other employees. They were never promoted. None of the other crew members wanted to be seen with them during off time.

They were always assigned the harder jobs on the crew. They very seldom received pay increases. And they were always blamed if something went wrong on a drilling site. Eddie told Grandpa that he knew more than enough qualified men who would be loyal to an owner who appreciated their hard work and knowledge.

I wondered how this item would be received by the others. Rex took the lead and said that he trusted Eddie and would gladly agree to that item. We spent some additional time creating the actual response to Eddie's demands. The completed job offer was well crafted, and everyone was sure Eddie would accept the position.

Rex brought up a few other things that needed to be addressed and when we finished, he asked Grandpa to meet with Eddie and ask him to come to a meeting at the ranch with all of us in attendance.

The rest of the week was spent with all of us meeting with a team of geologists who were experts in the drilling past and present on the Permian Basin. It was a giant learning curve for me. The others had spent more time on the subject while I was away at school.

Rex was trying to eliminate as much financial exposure as possible to increase potential profits and reduce potential losses. He could be described as a careful risk taker. This current opportunity was filled with risks and at the same time was full of rewards. His job was to balance them against each other to see which one might come out on top. That was the way he approached everything in his life. Doing things that way had served him well in the past and we were all hoping it would do so in the future.

During the next week, the girls were finalizing all the details for the wedding. The men were working on possible drill sights, cost to drill ourselves versus hiring the drilling, and waiting for Eddie to set the date for our meeting with him.

Many of the evenings were spent with all of us together at Rex and Abby's. Other times we were on our own with our families. Because the housing was the cabins, we usually ate at the Chuck Wagon before enjoying family time in one of the cabins. Being together was a wonderful benefit of working on the oil company plan. Maggie and I had been too young to ever be included in the planning and expansion stages of our family businesses.

That had all changed as we were all included in this new company. Each of us had a chance to offer our opinions on matters that were important for our futures. It was difficult to just have small talk because of the soon to happen wedding, and all the meetings about well drilling. We were all happy and in this together. That was the primary factor, together and happy.

The day of the wedding finally came. Rex had arranged for a local minister to conduct the wedding. There were no bridesmaids or groomsmen. Just Maggie and me standing before the pastor. He read a little passage from his notes about the importance of marriage and then he led us through our vows. When we were finished, he prayed for us and declared us husband and wife. Everyone cheered and I kissed my bride.

Maggie was beautiful in a simple dress that was not even a real wedding gown. She had told Abby that she did not want a wedding gown. Abby had bought her a very western dress that showed off her red hair and beautiful skin.

I wore new jeans and a western shirt with a bolo tie. We had a lot of pictures taken and then Rex announced that there was to be a small party for us down at the Chuck Wagon so that the ranch hands could wish us well.

This was without a doubt, the best day of my life. We spent a couple of hours with our friends from the ranch. Abby had ordered a wedding cake that was large enough for everyone to have plenty. We took a few more pictures. Rex made a toast to us and then announced that he was taking us up to see our new home.

Our families drove up to the house farthest from Rex and Abby's. It was perched on the same hill as the other three. Rex stood by the door as Maggie and I walked up. He was smiling as he handed us the keys to our new home. Then he said, "Tommy carry your new bride across the threshold into her new home." I did and when we entered, there was a huge banner across the entryway. It said, "Congratulations" I set Maggie down and as we looked around, we could see that the house was already furnished.

We learned that Grandma, Stephanie and Abby had to completely furnished our new home so that we only needed to move in and start our married life. Maggie started to cry. By the time she finished hugging everyone, I was getting emotional myself.

Grandma wanted to give us a tour. Every room was perfect. We did not need, food, towels, dishes or anything. It was all there! They had not forgotten anything. I could not find any words to say. Maggie had to thank everyone and do all the hugs again.

Rex finally spoke up, "Ok everyone, these kids need to be alone. Tommy, I do not want to see you in the office for a week." Then they all left. Maggie stood in the middle of our living room and was spinning around like she was at a ball or something. She was truly happy.

We walked through every room again and saw things we had not seen on the first trip with Grandma. Maggie said, "They did not forget anything". She was correct. We had it all. For a week, the only time we left our home was to go see Precious. It was like a dream. My life-long friend was now my wife, and we lived in our own new home working with our families for a man that we had learned to love as a big brother or uncle.

The week I was off from work, the other men had drafted a final proposal for Eddie. He was going to meet with us on my first day back, which was Monday. He did not show up until almost noon, so we had plenty of time to go over the last-minute details. Rex wanted Grandpa to conduct the meeting.

Eddie shook hands with everyone and took a seat. Grandpa started by saying that he had discussed most of the things that Eddie had on his list with our group. "I think it would be best if we talked about each item on your list so that we all understand what we are agreeing to and why they are important to you."

"Maybe before we do that you can give us some insight into your thinking when you came up with this list." Eddie was a lot like Grandpa. He measured his words before speaking. "Well Roy, I guess that I would say that I am happy with where I work. I have been there quite a while now and it would take something awfully good to make me change. Are there things that could be different at the company where I work? Sure, there are lots of them. Are they so terrible that I would quit? No."

" I have always thought about what my job might be like if I could make those changes. When Roy asked me to make a list of things I would want if I was to leave my job, I was able for the first time to put down all the ideas that have been running around in my head for the last 20 years ."

"So, ask me about any of them and I will do my best to explain why each one is important to me." Grandpa had the list in front of him and suggested that we start at the top and go through every item. "Ok Eddie, you want a ten-year contract." Eddie told us it was too much of a risk without having a guaranteed income for the next ten years.

Everyone agreed that any contract must include that provision. Grandpa said, "Item number 2, says you want to hire an all-Cherokee crew. What is your thinking on this request?" Eddie took a breath and said that life is hard for Indians in a world controlled by white people. He talked about how he had been looked down on by the white employers.

He talked about having to work much harder than the white employees just to keep his job. Then, because he was outworking the other men on the crew, he was treated poorly because they felt he was trying to show that they were not carrying their weight on the job. He said, " It is kind of darned if you do and darned if you don't."

"I know a dozen good oil field men who are Cherokee and all of them have experienced the same obstacles that I have. I would like to give them a good job where they do not need to worry about discrimination, and they are rewarded for their loyalty to their employer." Grandpa looked at Rex for an answer to that request.

Rex looked at Eddie for a long time before answering. "Eddie, what you ask and the reasons you ask it are very valid. If you think you can build a crew of good and qualified men who happen to be Cherokee, you will have my support. The one thing I will point out is that each of our employees regardless of their bloodline, will have to sign a contract that does not allow drugs or alcohol around our job site. And if they get in trouble in town for either of those contributing factors, they will be fired." Eddie said, "You do not have to worry about that. I would never hire anyone with those problems."

Rex suggested that we take a quick break and go to the Chuck Wagon for a snack. When we finished and returned to the office, Grandpa read the next item on the list. "It says here that you would want the drilling crew to share in the profits from each well. Can you explain that to us?" Eddie explained that year after year, he and the other crew members watched as the landowners and oil companies grew wealthy from their efforts. Eddie said that he felt it was only fair to allow them to share in the profits, from a well that they helped create.

Rex asked him what percentage he felt would be fair to the oil company and the employees. Eddie told him that his crew should receive 5% of the oil sales on every well. But if the well exceeded expectations, the percentage should go up to 10%. Rex told Eddie that he would need to run it by the attorney so it could be structured correctly. He said that he was not against that item, but it must be crafted carefully to protect all the parties.

We went over many other items on the list. Housing,

meals, insurance, vacation time, transportation and several other items were discussed without any problems.

Before we ended the meeting, Rex asked Eddie if he thought that he would be able to put a crew of his liking, together within 120 days. Eddie assured him that it would be possible. Rex told Eddie we would be in contact within seven days. Eddie thanked everyone and said he was looking forward to working with us in the future.

For the rest of the week, we met every day and worked on every aspect of getting the company moving. The first important thing to decide was whether we should hire a drilling company or form our own drilling company and do it ourselves. We spent two days going over all the costs that Andy and Grandpa had prepared. By late Thursday evening, the decision to move forward with our own drilling company was decided.

The lawyers had been busy and had created a preliminary employment contract with provisions for profit sharing. On Friday, we spent the whole day with them and made a few modifications to the contract. They agreed to have all the paperwork completed by Monday.

Rex had Grandpa contact Eddie and asked him to come a see us sometime in the next week. On Monday we spent most of the day going over the completed paperwork. We all agreed that it looked good. Rex wanted to put Eddie on the payroll as soon as possible. He felt that as we started to accumulate the equipment and supplies, that Eddie would be of great help in our planning.

He met with Eddie the next week. He was happy with the terms of the contact. Rex told him that we wanted him to join the company as soon as possible. He agreed to give his current employer the notice that he was moving on to a new company.

Rex told him that since he was not married that he could have housing in one of the cabins until we figured out the future. Eddie told Rex that he would get back to him when he found out what his employer expected from him after he gave his notice.

When our meeting was finished, Grandpa said he would give Eddie a little tour of the ranch. They were gone for several hours. Rex and was still going over a few things with Andy and me, when Grandpa came back.

Rex asked Grandpa how things were going with Eddie. It was a moment before grandpa could answer. "He is a good man, Rex. When I showed him the cabins, he was very pleased. For years he has been living in the temporary housing on drilling sites. He asked me if I felt that I had been well treated in our business dealings."

Rex asked what Grandpa had told him. "Well, I said that you were not a business partner, that you were family. He said Eddie looked at him and said, "Wado oganalii, donadagohvi." Rex said, "What does that mean?" Grandpa told him that it meant "Thank you my friend. You and I will see each other again." "So, is he happy? Is it a done deal?" Grandpa nodded and said, "It is a done deal"

And that is how Eddie Most became part of our oil company. He was a loyal friend and partner for the next 25 years. He helped us build our company. Never had a complaint. Was a fair and caring boss to his Cherokee crew. He lived a very humble lifestyle.

The next few months were very busy. Long hours and very few days off. We bought equipment and supplies, narrowed down where we would be drilling our first ten wells and started the permit process.

Having all the members of the company living on the ranch made it much easier to get things accomplished. Rex had sent Eddie to meet with a company who was actively involved in the fracking process. He spent a week there and came back with an understanding in how to modify traditional drilling to include fracturing.

By late September we were reaching a point where we could set a date to begin drilling on our first well. Eddie had his crew. We had taken delivery of a new drilling rig and large supply of drill pipe. Andy, Grandpa and Eddie had managed to accumulate all the support equipment and parts inventory that we would need on site. Our crew housing had arrived, and the trailers were all set in place. We had two large generators to supply power, and we were told that all our permits would be finalized in a matter of days.

We started drilling our first well on December 1, 1955. Every one of our family members was on site. Before the rig fired up, Rex wanted to pray. He asked God to bless all our hard work in getting to this day. He asked for safety of all our people. And lastly, he asked that God enable us to be good stewards of any oil that He chose to bless us with from this well.

When he was done praying, he had Abby break a bottle of apple juice on the drilling rig and say, "We christen you "The Future". Eddie and his crew moved onto the drill platform and the engine roared to life. We all watched as the first few pipes were added to the drill and our first well began.

Over the next weeks, there were many visits to that drilling site. Because the drill was new, we did not have many breakdowns. As the piles of drilling pipe grew smaller, we had hopes of finding oil. We were drilling an average of 150-450 feet per day. Our geology experts had told us to expect to find good oil anywhere from 6,500 – 8,000 feet deep.

As the days rolled by, our excitement grew. Our crew was getting to know the new equipment. Andy and Grandpa had ordered plenty of drill pipe. Two days before Christmas we hit oil. Our well was 7,400 feet deep. Rex met with the crew and told them that we needed to secure the site, and everyone would get a Christmas bonus and there would be no more work until Jan 2, 1956.

## CHAPTER THIRTY-FIVE

We had a wonderful Christmas. It was such a blessing to have all our family members living in the same place. Rex was true to his word, and a new drilling company was formed. It was named Cherokee Drilling Company. Each member of the crew was to have a percentage of profits of the from each well. The rest of us worked on the same principles but with different percentages.

Each person who was getting any percentage was to be paid monthly, with one half their share paid into an investment account and the other half paid in cash . Rex wanted each person to be building some personal wealth in the investment fund.

By the time the crew received their first profit sharing check, we had moved the drilling rig and were several thousand feet down on our second well.

The first well which Rex had named Abby No. 1, was now producing 980 barrels per day. It was paying off some of the investment that we had made to start our own company. Everyone was pleased with our progress. The crew members were very happy, and Eddie acted like it was just plain fun and not a real job.

He was seeing dreams come true that were normally impossible for a working man to ever experience. His crew was more than a working group, of men. They were his family. He was responsible for them and their futures. He took his job very seriously and every one of them looked up to him with respect and admiration.

He had already started contacting more Cherokee people to get them interested in being part of our company. Rex met with Eddie at least once a week and talked over every aspect of our little company. During one meeting Eddie told

Rex he would like to be able to bring on another crew of less experienced workers.

He wanted to train them to do things his way. He felt that they would be able to offer a second experienced crew by the end of the year. This second crew could run a second drilling rig. That plan was music to Rex.

He gave his full support to Eddie and within a couple of months the new recruits started showing up in our camp. Abby No. 2, came in and when it was put into production, we were getting another 1,050 barrels a day. That brought our total to a little over two thousand barrels a day. Things were coming along better than we had projected.

By mid-summer 1956, Eddie told Rex that we needed to start considering purchasing the second rig. His new crew was coming along faster than he had originally planned. Eddie felt that they would be ready within six weeks. Andy and Grandpa were turned loose to buy and second rig and all the necessary support equipment. Eddie had already chosen a new foreman for the second rig.

By November, our second drilling rig was in operation. Our second crew was doing a great job. Eddie had trained them well. Rex was in his glory. He had two producing wells and two drilling rigs. His master plan was coming true right before his eyes.

By the end of January 1957, we had three producing wells and the fourth was almost finished. Abby No. 3 was our best well to date. It was producing 1,100 barrels per day. But nothing would have prepared us for what happened when Abby No. 4 came into production. After one week of pumping, that well leveled off at 1,370 gallons per day.

We did not have a single well that was more than 9,000 feet deep. Our drilling cost to date, was under what we had projected. Our company was doing a great job. The next four drill sites were already staked out and 1957 was shaping up to be a huge winner.

Our oil company, WSM was growing at a speed that even Rex was surprised. Our wells were doing more than expected and the profits were coming in at a rate that could not have been anticipated.

There were companies of all sizes trying to meet with Rex and see if there was a way for them to get some of the action. Rex always gave them the same answer. "We do not sell leases on our land, and we do not need partners." It was not what they wanted to hear. The Walker empire kept on growing. It was just as Rex had predicted, a great adventure. Grandpa had never seemed so alive. Andy had become a right-hand man to Rex

I was being trained in every aspect of the oil business and the financial machine that was needed to keep it running. It was far from being boring. I loved what I was doing and would not have traded it for any other opportunity.

Grandma, Stephanie, Maggie and Abby were busy with girl stuff. They worked with the charities that we supported. Maggie was spending time with Adam several days each week. Precious was at a point in her training that the rider was less important. Precious knew her job and she loved every minute when she was working cattle. Adam had suggested that Maggie should ride her in a couple of big rodeos so she could have some competition experience.

A schedule of several venues was finalized, and Adam and Maggie were getting ready to hit the road. It was a big step for Maggie, but Adam was always there to reassure her that she and Precious were up to the experience. Rex had planned the schedule so we could fly to a couple of the competitions and watch Maggie and her horse compete.

In the fall of 1957, we hit our first well that was not worth pumping. Rex was very positive about the whole thing. He constantly reminded us that not every well was a gusher. Overall, we were one of the most profitable companies doing business in our part of the state.

By the summer of 1958, Maggie had earned three titles on her horse and had received a nice amount of prize money. Precious was becoming a name in the world of cutting horses. It was just one more thing that was making Walker Ranch a household name.

During 1958, Cherokee Drilling Company had completed an additional 10 wells. Three of those were of no value. Our record of winners exceeded our expectations. Rex wanted to try and have each of our two drilling rigs produce 6 wells In 1959. Eddie assured him that the crews would try to meet the goal.

Word gets around in the oil business. Eddie had men showing up almost every day asking for a job. They had heard about the profit sharing and good working conditions. The way we did business with our workers was not as popular with the other companies as it was with people looking for a job.

Over time there was pressure on oil industry leaders to take better care of their employees. I choose to believe that our business model had some influence on creating a better working environment for oil workers throughout Texas.

By July of 1959, our goal of 12 wells for the year, was on schedule. Each rig had completed 3 wells. One was a real winner at 1,605 barrels per day. Three more were average and two were of no value.

Maggie had won several more events and we were now having interested parties making offers to purchase Precious. Maggie always told them that Precious was not for sale. Stephanie had started going on the road with

Maggie and she seemed to be enjoying the whole experience. Grandma was busy helping Abby and those two were growing closer by the day. Our whole operation was just one large family doing business in Texas and Oklahoma.

Rex had allowed the girls to make trips to Oklahoma to check on business in Afton. Rex, and Everett, our ranch foreman, would fly up and check on the feedlot and cattle in Afton, at least once per quarter. Our people there were totally able to run that operation with just an occasional visit from Rex. It was a blessing to have employees of that quality.

Christmas of 1959 was one of those that you will never forget. Rex took Grandpa and Andy with him to Afton. They had the traditional celebration and passed out the bonuses. When they arrived back at the ranch, Rex wanted to include the oil drilling crews in our Christmas dinner for the ranch employees. It was a wonderful celebration, and we were able to honor and reward all those wonderful people who made this place called Walker Ranch such a success.

The bulls that Rex and I had purchased and introduced to our herds were doing their job. Walker ranch produced some of the best beef that money could buy. The 12 wells that Rex had requested, were all completed by December of 1959. The last six had 5 winners and one loser. It had been a good year. But it was going to get better. On New Years Eve, we were all together at Rex and Abby's. We were talking about the great year we had just experienced. The girls were in the kitchen making some snacks.

Suddenly, there was a lot of noise coming from their area. The men did not know what to think so we just stopped talking and were listening. Eventually they all came out. Grandma was crying. Stephanie was wiping her eyes with a cloth. Abby was holding Maggie's hand and guiding her into the great room where we were sitting.

None of us could have possibly guessed what had prompted this reaction among our women. Grandpa was the first to speak up. "What is going on? Has someone died?" Grandma was unable to speak. Finally, Stephanie looked at Andy and said, "Andy you are going to be a grandfather."

I looked at Maggie and she just nodded. I jumped up and ran over to hug and kiss her. It goes without saying, the rest of the evening was a big party. The love in the room for Maggie and her announcement was something that touched everyone in the room. Rex could barely get out the words, "Well I guess we are going to have to find a suitable filly for our new member of this family."

Our lives would never be the same. No one knew what that baby would mean to each person in that room. When everyone finally settled down, the main question was, when will this little one meet its new family? Maggie was given the opportunity to share with us what had been happening. I guess she had thought she might be pregnant but wanted to make sure. On her own she contacted a doctor in town and had it verified. She was pregnant and the baby should show up in late July.

I guess that It never occurred to me that we were about to become parents. You always know that kids are in the future. When the future becomes the present, it gets real very quickly. For many days Maggie and I enjoyed the fact that a baby was coming. As the weeks went by, I could swear that she looked prettier than she ever had.

There were so many people who loved Maggie. They were constantly trying to make life easier for her. Maggie had to remind them that having a baby was a normal thing women did every day. It was not life threatening and she would be able to get through it.

She did however agree to not go on the road with Precious. Andy was able to ride her and did very well that next year. Life moved on but each of us was counting the days and weeks before the baby came.

By the first week of July 1960, our crews had drilled an additional 7 wells. Four of them were producing over one thousand barrels per day. One was around 650 and the other two were not viable.

When Maggie had her checkup in March, the doctor told her that he thought he had detected two heart beats. I guess he was right because at the end of April, he told us we were going to have twins. Maggie was overjoyed. Every member of our "family" was excited. We did not know if they were going to be girls or boys or one of each. At almost every meeting of the families, the subject would turn to possible names.

Maggie and I had come up with our choices, but we were keeping them to ourselves. The big day came on July 27th. Maggie was being taken care of by the girls. They attended to her as if she was a queen. I was at work with Rex that morning. He got a call from Abby and that she and Grandma were leaving for town with our nurse from the clinic on the ranch. Stephanie and Andy were leaving soon with some things that might be needed at the hospital. Rex said, "Well Tommy we better go see what Maggie has been working on for the last 9 months."

We picked up Grandpa and made the trip to town. It was a five hour wait before Maggie produced the first of two little girls. They were both tiny, very fine white complexion, And a wild mop of flaming red hair. Our names were announced by Maggie to the staff as they cared for each of the babies. The first born was Molly Ann Meadors and the second was Mandy Sue Meadors.

My life with three girls all with names starting with the letter M would never be the same again. But it was the second happiest day of my life. Those little ones had no idea how much they were loved by our families. They had two sets of "grandparents" and one set of "great grandparents".

Maggie recovered well and all three of them were headed home in a few days. One of the extra bedrooms in our home had been converted into a nursery with two matching cribs. The doctor had tied a little ribbon around Molly's wrist so we would be able to tell them apart for the first few days.

Grandma, Stephanie and Abby took turns spending the night with us and helping Maggie care for two babies. Every time that I tried to help, I was pushed aside and told to go find something to do. All of us men kept busy and decided that the women knew what they were doing.

By the time the babies were a month old, Maggie was feeling much better. Grandma had discovered a little freckle on Molly's left ear. And that became the way we could tell them apart. They were about as identical as it was possible to be. But time would show that they had very different personalities.

## CHAPTER THIRTY-SIX

Christmas 1960 was special in many ways. Our oil company was doing very well. We now had 35 good producing wells and a couple of average ones. Grandpa and Andy accompanied Rex to Afton for the annual Christmas celebration at our plants. When they returned, we repeated the process with the ranch hands and the drilling crews.

Our people were happy. Our companies were doing better than expected. And the babies were bringing happiness to all of us. Maggie had to ask the other women to stop giving so many gifts to the girls. Every one of us loved those girls so much that it was difficult to know when to stop showering them with things we thought they needed.

They were approaching six months in age and little things were starting to show their individual personalities. Molly was more independent. After she had eaten, she wanted to play with something. When Mandy finished, she was content to just be held. They were a joy to every one of us. Rex was the most fun to watch as he had never been around babies. He held them and played with them as though they were breakable.

The price of oil had risen to about three dollars a barrel. We were doing very well. 1960 a new organization was formed. OPEC stood for the Organization of Petroleum Exporting Countries. It would influence the oil prices for many decades into the future. For now, things were stable. The biggest regulator of pricing for us was the Texas Railroad Commission. They regulated the amount of oil production in our area.

Grandpa had his 70th birthday this year. Rex made sure it was special. We had a big barbeque, and everyone was invited to his milestone party. Grandma produced a few photographs of them when they were young. It was a wonderful time.

We celebrated Grandma's 70th in 1961 along with the twins first birthday. Maggie had gone crazy with preparations. The cooks at the Chuck Wagon had made a huge cake and everyone was invited. Molly and Mandy were the only two children on the ranch who had red hair. Precious knew they were something special from the first day she got to see them.

I think the red hair was the give-away. She had nuzzled them and smelled their hair. It was a ritual that had been repeated dozens of times over the last year. When the birthday party was over and lots of pictures had been taken, Maggie told the girls that she had a special surprise.

We took them down to see Precious, but this time they were allowed to sit on her back as Maggie slowly walked Precious around the corral. They were very excited and laughed and giggled the whole time. That was the start of their journey with horses. Rex had big plans which would assure that the girls would always have horses. After all, he looked at those twins as his own grandchildren.

Oil prices remained stable and in 1961 we added seven more good wells to our company. Eddie was wanting Rex to add a third drilling crew and another new drill. He assured Rex that he had the manpower. several months, Rex put off making any decision, one way or the other.

At the twins second birthday party, Rex told Eddie that he was content with having two rigs and wanted to continue drilling at our current pace. I think Eddie was a little disappointed at first, but he trusted the opinion of his boss who was making him a very rich man.

Maggie had started back running Precious on the circuit again. It brought her so much pleasure. Stephanie would accompany her on these trips and watch the twins. Those two little people knew more about horses than any one of that age should. It never ceased to amaze me as I would see one or the other of them leading around a full-grown horse.

On their third birthday, Rex allowed them to do what he had Maggie do all those years before. They each got to walk through the stables and pick out their own horse. Then they did as their mother had done before them, name them. Molly named her choice Flower and Mandy called hers Bonnet. I did not question the names. Grandpa Rex as they now called him, gave his total approval to the names.

No two children ever had as much attention as my two girls. There were so many people in their lives who loved them and wanted to constantly be with them. Grandpa Rex took them on plane rides. Grandma Ida let them help her bake and make quilts. Grandma Abby took them shopping and made sure they always had new outfits to wear.

Adam let them spend time with him when he was training Precious and working with their horses. Grandpa Andy and Grandpa Roy had to fight to get their time which was usually leftovers from the women. They would take the girls on short exploring trips around the ranch. Grandpa even played catch with them as he would have done with a boy. One could say that they had perfect lives. Endless possibilities provided by those who loved them.

Maggie and Adam had been discussing how much longer that Precious could keep competing. Adam felt that the following year should be her last. Maggie agreed and made plans to show her as many times as possible before her retirement.

By Thanksgiving 1964, our company had 59 producing wells. The price of oil was steady at around $3.10 per barrel. As was his custom, Rex met with us to discuss the plan for the following year. He had decided to keep the current drilling schedule. Eddie agreed that it would be very comfortable for the crews. The companies and feed lot in Afton were doing well, and Rex did not want to make any changes. The profits from our oil company were continuing to grow even with our girls donating to several charities.

Rex informed us that he wanted to enlarge our little school on the ranch. He wanted the children who lived on the ranch to have the best education possible and not have to go to town as they got older.

He asked Andy and Grandpa to do the research on how and when we could get this project started. When he was describing what he wanted to do, I could not help but think it was mostly because of Molly and Mandy. Typical Rex, he got his wish and by September of 1965, our little school became a new big school. We had a dedication party. It was named Walker Ranch School. It had 12 grades and Rex had Andy and Grandpa hire the best teachers that money could buy.

Over the next decades, that school produced crop after crop of well taught students. Another Rex dream had come to be a reality. In September 1966, our twins entered first grade. Maggie finished her last season showing Precious. Rex purchased his first jet aircraft. Abby became the president of the Texas Women's Philanthropy Association. Rex was voted Texas Rancher of the Year.

Grandma published another cookbook. Grandpa started building a dock into the pond at the special place where I had proposed to Maggie. It would be many years before I realized the significance of that project. I spent most of my days with Rex. He was grooming me to take over the ranch at some point in the future. We never talked about it, but it was too obvious.

In 1967, Maggie bred Precious to a stallion who had also been an award-winning cutting horse. As soon as the word got around that Precious had been successfully bred, there were several offers to purchase the foal. Maggie had not even considered selling any of the offspring from Precious. They would remain on the ranch and for those who showed promise, there would be the usual training process.

The ranch foreman Everett had told Rex he was retiring. He had served the ranch faithfully for more than three decades. Rex had a big party for him. Adam was now spending his time with Flower and Bonnet. Grandma was teaching the twins how to cook. Mandy loved it. Molly could have cared less. She only went along with the lessons to please Grandma. Those two girls may have been twins but they sure had their differences. Mandy loved school. She loved being with the other students. Reading was her passion. Molly just wanted to get the school day over with so she could go play with horses. Learning schoolwork was of no interest to her.

They were both very bright. By the time they had finished first grade, they were reading well above their age level. I think Molly was Rex's favorite. She was a risk taker and a wild thing. But he loved Mandy's gentle way with people. Her constant concern for others was a wonderful quality.

Grandpa disappeared several times each week. He was gone all day and returned just before dinner. I knew he was working on his project by the lake. Rex was respectful of Grandpa's age and did not try and over work him on the ranch plans. When he needed something to do, Grandpa would ask Rex for a new assignment. Rex always had new things going so there was a never-ending list for Grandpa to choose from.

Andy was busy every day. Between the oil company and

watching over the things in Afton, he was never bored. Stephanie was working with Abby most of the time. The charities always needed managing. Grandma had started teaching a cooking class at the school twice a week.

Maggie spent most of her time helping Adam with horses. What a crazy few years it had been on the Walker Ranch. When Maggie and I were young kids in Afton, there is no possible way we could have imagined this life we were living. Grandpa always said that God never made mistakes. His plan for us was beyond our imaginations.

Maggie still called me Gule and always wanted to protect me. She was my best friend and greatest supporter. My life could not have been more perfect.

Grandpa was spending more time with Molly. He loved spending time with her. For more than a year he had been playing ball with her. He even nailed an old tire to a fence so she could learn to pitch the way that he had taught me. I watched from a distance many times. She was good.

For a girl of not quite seven years old, she could throw like any boy and better than some. Grandpa must have seen her potential because he worked with her almost every day for an hour or so.

It did not take long for Rex to become aware that Molly loved playing baseball with Grandpa. He had Grandpa come to the office on afternoon. "Roy, we have this school here, but these kids are missing out on playing sports. What do you think about us building a baseball field and you can start teaching these kids how to play?" Rex told me later that Grandpa sat there for a long time before he answered. "Rex, there is nothing that I would rather do."

In Rex fashion, within a few months we had a very professionally prepared baseball field. Grandpa made sure that they were provided with baseballs, bats, gloves and anything else he asked for.

It changed his life. He had an opportunity to be a part of something he had loved his entire life. I looked at him each day and what I saw a younger man full of energy and purpose. I am not sure that Molly was very happy about sharing her grandpa time with the other kids. She went along with the plan to make him happy.

For several weeks, Grandpa allowed any of the school kids to come out to the field and do some simple drills with him. Throwing, hitting, running and learning the rules were a few of the things he was teaching them.

I would sometimes stop by and watch. It took me back to those special times when Grandpa and I practiced after we had finished our chores. And now, I was watching my own little daughter enjoy those times with him.

In 1967, Precious had a little filly. The twins were there to watch her give birth. Maggie asked them what name they wanted to give the little horse. After some discussion they settled on Usdi. Grandpa had taught them a lot of Cherokee words, and they knew that Usdi meant baby or little. So Usdi became her name.

Over the next few months, both girls visited Usdi each day. It reminded me how their mother visited Precious every day. That little horse drew my twins closer than they had ever been. Usdi was part of their world from the day they saw her born. It was a magical relationship. They smothered that little horse with kisses and hugs. She was groomed and brushed so much that I thought her hair would fall out.

It made Maggie cry on more than one occasion as she watched them loving on their Usdi. Rex put out the word that Precious and her baby were to be treated like royalty. Maggie was struggling with retiring Precious. Each time

Maggie showed up at her stall, it was very apparent that Precious wanted to go work. It was more than Maggie was able to handle so every so often she would saddle up Precious and together they would go herd some cattle.

Abby and Rex loved their "granddaughters" and spent as much time as possible with them. What Rex had done for us was beyond my dreams. I think in return we had done something for Rex and Abby. We had allowed them to have a family and to surround themselves with people who loved them. Grandpa said it was God's plan.

## CHAPTER THIRTY- SEVEN

1968 passed without any problems. Grandpa was creating two teams at the school who were now playing baseball against each other. He was in his glory. Molly was one of his most dedicated players. Mandy was becoming a real homemaker. She spent most of her time away from school with Usdi and helping Grandma quilt and learning to bake.

The oil price was steady, and we added eight good producing wells during the year. Adam and Maggie were training Flower and Bonnet. Surprisingly they both showed promise.

As Rex and I were planning the usual Christmas programs for the ending of the year, I could see some changes in the man who made all our lives so much better. He was a bit distant and did not seem as driven as he usually did. I tried several times to approach him with my observations, but something interrupted us each time.

1969 was one of our best years in our oil drilling . We hit 9 good wells out of nine tries. That was pretty much never heard of results. The bonus and profit sharing was making most of our crew members quite wealthy. Rex and I had the conversation about what we might do when some of these workers felt that they need not work anymore because they had so much money in their profit-sharing accounts.

We agreed that we needed to address the concern with Eddie. We scheduled a meeting with him for the first week of May. At that meeting Rex told Eddie our concerns. Eddie told us that there were a few of the older guys who were talking about wanting to go enjoy their money while they still had their health. Eddie said he had even thought about it himself.

Rex asked him what he felt was the best way to prepare for the inevitable. Eddie told him that there would be no problem replacing our workers. He had accumulated a waiting list of qualified men to take their places. Rex asked Eddie for a commitment to stay around for a least another year and get replacement men in place for those wishing to retire. Eddie gave Rex his word that he would make it happen.

Every man on the drilling crews got a statement each month with their share of the profit and the balance that was growing by earning interest. Most of the workers were becoming quite wealthy as compared to other working people. It was no wonder that they would like to enjoy some of that money and go do things that they had not be able to do while working.

Rex asked Eddie to create a list of those men who were close to deciding to retire. When he had the list, we would meet again and put a new plan in place. When the meeting was over, I asked Rex if we could talk a bit. "What's on your mind Tommy?" I told him that I had seen some change in him over the last few months. He sat there for some time before answering. "I did not know it was obvious." I told him that it was concerning to me and if he felt like talking about it, that I would be happy to listen.

"Tommy, it seems that I have reached a point where everything that I had dreamed of doing has become a reality. The oil business, our businesses in Afton, our new breeding program for our cattle. Then our school and our baseball program. We are drilling our own wells. Precious has put the Walker Ranch out there as a real producer of award-winning cutting horses. It just seems that I am running out of things to do. What is my purpose now?"

I was sitting here looking at a man whose ideas had

changed so many lives. He was the most talented man I had ever met. Yet here he was, depressed over his life. I had no words. I was so deeply moved by what he had told me.

I got up from my chair and walked around the desk a grabbed him and held him in my arms. We both held each other and shed a few tears. After a few minutes I let go and when I could finally speak, I said "Maybe you should stop making big plans and enjoy each day with those who love you. You are 62 years old. You do not need any more money or projects. You need us, your family. Take your beautiful wife Abby for a horseback ride on your ranch. Go down and see what Grandpa has been doing by the lake.

Take flying lessons so you can fly your own jet. Go visit the Cherokee Nation and let people meet the man who was willing to hire only Cherokee men to run his drilling business. Take your granddaughters with you and Abby. Go on a trip somewhere you have never been. Build some memories that will keep you warm in winter when you grow older.

Allow some of us to do something for you. Not because we feel we owe you but because we love you." I do not know if Rex really understood that we did love him. He was trying. It was on his face. He was trying.

When I finished, he sat looking down at his desk. When he spoke, it was slow and not very steady. "Thank you Tommy. Thank you for loving me." That is all he said. He just got up and walked out of the room.

The next day Abby called me and asked me to stop by her house after lunch. I finished up what I was doing and went up to see her. She met me at the door and gave me a hug. "Tommy, I wanted to ask you what happened with you and Rex yesterday. When he came home, he was very quiet and when I went to tell him dinner was ready, I could see tears running down his face. He hardly said anything during dinner. When I asked him what was wrong, he asked me if I was happy with my life. I assured him that I was very happy. He told me that you had told him that you loved him." Then she began to cry. Now, I was about to cry. "He told me that he did not know what to do with his life.

That everything he had planned had come true. I told him to let us love him and to enjoy what he has now and to make memories for the future and that there was no need to plan for any more big things in his life."

"Tommy whatever you said made a huge impression on him. He asked me to make plans for a trip somewhere with just the twins and us. He told me he wanted to see some big ranches in Australia. He even told me he might take flying lessons."

I could not believe what I was hearing. Rex never indicated that he was taking to heart the things that I had said. He must have been listening because Abby repeated things that I had told him. I will say that the years following that private conversation between two men who loved each other, changed us both for the rest of our lives.

We talked a bit longer. Then Abby wiped her face and said that she needed to get dinner ready. She threw her arms around me and kissed my cheek. "Thank you Tommy. I love you so much."

By the spring of 1970, Eddie had produced a list of everyone who worked on our drilling crew. Each member had expressed their plan of continuing to work or to retire. Eddie gave us a promise to work five additional years. Rex was very happy. Any changes in personnel would be done before Eddie retired.

Adam and Maggie were working with Usdi who was approaching the age of two. Flowers and Bonnet were getting more advanced training. Both were doing very well.

Precious was not going to have another foal this spring. Grandpa had been contacted by a school in San Angelo who wanted to play a baseball game against our players. He was more excited than I had seen him in years.

Abby was finishing up the plans for a visit to Australia the coming summer. She and Rex were taking the twins and planned on being gone most of the summer. You could have knocked the rest of us over with a feather. Rex was leaving and left us in charge of everything. He really was changing. He seemed happier. It was if he was enjoying one day at a time.

He had even started investigating flying school. It was not uncommon for him to show up and watch baseball practice being conducted by Grandpa. Some days he would saddle up and ride out on the ranch. Other days, he and Abby would take a picnic and go together. He was even seen roaming around the school and talking to students.

Abby had been able to get Rex to speak at some of the charity dinners. That was something he would have never agreed to in the past. Rex had Andy handle most of the employee business with Eddie. We were going to lose a few men who had been with us for many years. Eddie had replacements for each of them. Production would not be interrupted on either of our rigs.

As summer approached, Molly and Mandy could hardly contain their excitement about their trip to Australia. Abby had planned out every day of their vacation. Rex did not need to do anything. For the first time in his adult life, others were taking care of him. As the days passed by, he seemed to be enjoying the new experience.

Grandpa finally accepted a date to have our school play the one in town. Most of the ranch folks drove to town to see the game. We ended up losing 7 to 5. No one seemed to care. Our team had played a game with another school. Grandpa was very happy he was part of the team and experiencing a game that had real players. Our team wanted Grandpa to set up another game with a different school. Baseball was back in his life after all those years had gone by.

When the last day of school was over, Maggie was helping the girls pack for the trip that was only two days away. We would miss them terribly, but this was a wonderful opportunity for them, and it was even more special to make the trip with Abby and Rex.

Maggie had purchased a new camera for each of them along with a large supply of film. The plane flew them to Dallas where they would catch a flight to New York and then to Australia. We met Abby and Rex at the plane. There were hugs and kisses before they boarded the plane. It was hard for Maggie and I to see them go. They had never even spent a night away from the ranch and this was going to be several weeks.

Rex told me if there was something very important to leave a message at his hotel and he would call back. Then he looked at me and said, "Thank you Tommy, I am liking my new life. You will not need to call me. You and Andy can handle whatever comes around. We will take good care of the twins." Then he gave me a hug and got on the plane. We waited until the plane was off the ground and headed back to a ranch that would be missing four people who we dearly loved.

We got a phone call once a week or so. It was an update from the girls about what they were doing on the trip. Rex never once asked about the ranch. He was a man who was making peace with life.

The four of them returned one week before the girls

started school. They would attend the fifth grade this fall. Beef prices were up and oil was as high as $3.20 per barrel. The ranch was doing well in each of our enterprises.

Usdi had already won several competitions. Flower and Bonnet would be entered into competition this fall. Maggie lost herself in her horses. Precious was bred again in the fall and she was doing well. We had so much good luck with Usdi that we bred her to the same stallion. Precious had her second foal. It was another filly and the girls named her Sunrise.

When Rex was ready to get back in charge of things, he approached his position with a new way of doing things. It was now about maintaining our companies and doing the best job with what we had. He was no longer driven to grow or expand. After being home for the holidays, he told us that he and Abby were leaving in the spring to go see the agriculture in several South American countries.

For the next couple of years, that is what he and Abby did. They took trips to many places in the world. He was still interested in the farming and ranching in each place that they visited.

In 1973 there was a major change in the oil business. It was called an embargo. The Arab countries cut off oil production and sales to many countries because they supported Israel in the Yom Kippur War. By October the price was $5.12 per barrel and by December it was $11.65. We were getting almost four times the price for our oil.

Rex felt that this would be a temporary thing and did not to invest in ramping up our drilling beyond what we were on schedule to do for the next two years. Our profits soared. Wells that came in as good producers were paid off in weeks not months. It was unreal.

Eddie was doing his job as usual. All Cherokee crews who

could be counted on to do their jobs well. Our crews were the envy of all the other production companies in Texas. We had good men because we paid them well and treated them fairly. They were respected and not looked down on because they were Indians. Eddie was the one who made the plan and it was proving every day to be a model of how you treat people.

In May of 1974, Grandpa did not show up for breakfast. He was in bed, but he was no longer with us. He did not quite make it to his 85$^{th}$ birthday. He was the first loved one that I had ever lost.

It turned my world upside down. Andy and Grandma had to take care of the details. I was of no use to anybody. I went down to the stables and found Precious. I poured out my heart to her. I hugged her and cried.

Horses can sense your emotions. She just stood there and let me hold her and cry. When I could no longer stand up. I laid down in the stall and fell asleep. I was awakened by moisture on my face. Precious was breathing on my face to see if I was ok. I looked up and told her that I would be ok.

I went back to my house. Maggie met me at the door. "Where have you been?" I told her that I had been with Precious. She grabbed me and held me tight. "I am so sorry Gule". We both stood there crying. "You need to go see Grandma." That quarter mile to her house seemed like 50 miles. I had no idea what I would say to her.

When I opened the door, Abby and Stephanie were with her. She looked older than a few days ago. She had a cloth in her lap to wipe away the tears. She looked up as I entered the room. She held out her arms and I reached for her. We just held each other for a long time. She felt small and frail. I did not want to think about losing her and then having either of them.

When we finally let go of each other, I could not speak. I just shook my head and looked down at the floor. She said, "Gule, donadagohvi." I knew what that meant. It was what the Cherokee said instead of goodbye. It meant "we will see each other again". She looked at me and just said, "My oak is gone but I still have my acorn."

She went into the bedroom and returned with an envelope addressed to me. She wanted me to open it. Tucked inside was one sheet of paper. Grandpa had carefully written a letter to me.

"Gule, I have gone to see the grandfathers and my God. I want you to see what I have prepared by your special place. I want to be buried there. Take care of Grandma. Wado Gule, donadagohvi." He was using his last words to thank me. I was the one that needed to thank him. I handed the letter to Grandma and went outside to be alone. A few days ago, my life was so full and beautiful and now this loss had torn my heart to pieces.

No one could console me. I was imprisoned by my grief. Maggie left me alone. The twins were doing their own crying. Thankfully the others were tending to Grandma.

We honored his wishes and buried Grandpa by the little lake and under the big cottonwood tree. I know why he made the place. It reminded him of our special place by the lake in Afton. He had completed a dock and a picnic table. I would visit it hundreds of times over my lifetime. Rex had a tombstone made that said, "Here lies Roy Meadors, a man who loved God and who treated every person with respect, we will miss him."

After Grandpa was gone it was easy to see that Grandma had lost her will to live. She stopped teaching at the school. The twins would spend the night with her quite often. She seemed to enjoy that very much. Every week or so she would ask me to take her out to his grave. It became a special journey for us. I treasure those times. They did not last long enough. She passed away almost a year from the time we lost Grandpa.

She was buried next to her oak. Rex ordered another tombstone. "Here lies Ida Meadors a woman after God's own heart." I had two reasons to visit this spot and when life was getting me down, I made my way out there to be with them.

We had lost one of our families. With time, their passing became less painful. It just never quite went away. The price of oil continued to climb. Rex was giving more in bonuses to all our employees, even those who did not work in the oil fields.

That spring we lost Precious. She died quietly in her stall one evening with Maggie holding her head on her lap. It was the end of a perfect relationship between Maggie and her horse. It had lasted for two decades. They had grown up together. It was not unexpected but was still a huge loss to Maggie. There would never be another Precious. After her death, Maggie seemed to lose some of her desire to train cutting horses. She asked Rex if we could bury her near Grandma and Grandpa at our special place. Rex had her tombstone made and it read "Precious was never a horse she was part of our family."

Rex and Abby were still taking trips. During one of our meetings when he was home, I asked him what he wanted to do with Grandma's house. He looked straight at me and said, "It is Grandma's house and no one is going to live there until one of the twins gets married." The subject was never brought up again.

At the beginning of March in 1976, Rex asked what we were going to do about the girls and college. They were just a little younger than when he made plans to send Maggie and me to A&M. I told him he needed to talk with them and see what they wanted to do. He was good with that.

Molly had started attending the cutting competitions with her mom and Adam. Maggie rode Usdi for the last time. Flower and Bonnet had already been retired and they were both bred to good quality stallions.

Mandy was more interested in doctoring every single animal on the ranch. It did not matter if it was a mouse, a bird, a cow, or a horse. Her love was making them feel better. She would go along with her sister and mom on some of the trips, but she did not really care what horse won. It was all about the animals.

For more than a year Mandy had been spending time with Arnie Wilson the main veterinarian on the ranch. She helped him with vaccinations, worming, minor surgeries and preventative medicine. Arnie knew that the girls were talking about school. He took it upon himself to have a meeting with Rex about Mandy's future. Rex had no idea that Mandy was doing all those things on the ranch.

Just prior to Easter Vacation, Rex asked the twins to come to his office. They knew something was up because we lived so close to their house that they could have just dropped by after dinner for a talk.

Rex told them that it was time for them to give consideration for their future. He expressed his desire that they attend A&M. He reminded them that he had sent their parents to A&M when they were about the same age as the girls are now. He asked each one what they would like to study in school. Molly hesitated so Mandy jumped in and told Rex that she would like to study animal science and maybe even become a veterinarian.

Molly waited for Mandy to finish and then she told Rex that she was not sure what she wanted to study. She said that she would like to try out for the girl's baseball team. "I do not know what to study. I like my ranch life and riding the horses in competition. What do you think I should learn that I do not already understand?"

Rex told her that ranching was a business. "Every successful ranch has someone watching over costs. Managing the money is as important as managing your horses and cattle. Maybe you should take some business classes, and you can see why that part of ranching is so important."

Molly was listening to everything he said. When he was finished, she said, "Can I start that direction and change later if it does not suit me?" He assured her that it would be fine. In just one meeting it was finalized. Rex said he would have someone start making the plans. "Molly, about the baseball thing, that one is on you. We will get you in touch with the right people, but you will have to seal the deal"

The girls told their mom and me at dinner that evening. When we went to bed, Maggie leaned over and said "Well Gule, we are getting older. Our little girls are following the path we took all those years ago." Long after Maggie had fallen asleep, I was laying there thinking about all that had happened in our lives. I was hoping Molly would make the baseball team. It would have made Grandpa so proud.

In the fall of 1976, the girls left the ranch in one of our trucks and drove down to the school. Every year there were a few of the ranch kids attending A&M. The twins would have friends there to show them around. Molly had been in touch with one of the coaches and was going to have an interview with him before she would be able to attend any tryouts or practice.

Oil prices were still climbing, and our production numbers were very stable. Rex was turning 70 years old next year and Abby had been planning a special trip for them.

Cattle prices were firm and our horse training was getting some world-wide attention. There were wealthy people from the Middle East who wanted to buy our horses. Maggie always gave the same answer. "We do not sell our horses".

This was hard for them to understand because in most places around the globe, there was always a number that would convince you to sell whatever they were trying to buy. They had never met anyone like Maggie. The fact that they could not buy our horses only made them try harder.

The girls came home for Christmas, and we had a wonderful time. Rex and Andy were back from Afton, and we had the annual Christmas party for the ranch and oil company personnel. At the party, Eddie Most, told us that he would work through 1977 and that would be his last year. He assured Rex that he was training his replacement and that they would not even miss him when he was gone.

Rex was sad to think about Eddie leaving but was happy that he would be able to retire a very wealthy man. Rex told Abby that he did not want to take any trips in 1977. He wanted to be around the ranch and see the twins while they were on the ranch for the summer. Abby being Abby, she changed gears and started planning an epic 70th birthday for Rex.

Andy and I had pretty much been running the ranch, the oil company and we helped Eddie with the drilling plans. It kept us busy. Andy had the additional responsibility of the two companies in Afton. It was going to be a lot easier with Rex being around in 1977.

Easter Vacation was a great time with the girls. Molly had received permission to attend spring training and try and make the team. She was probably the only player there who had not been a star in high school. Her advantage was that her grandpa had trained her and she had been practicing since she was 5 years old.

While she was home, baseball was all she talked about. I was concerned about her feelings if she did not make the team. Both girls were happy with what they were learning at school. It was a wonderful visit.

We had dinner with Rex and Abby the night after the girls left. Everyone had a chance to share what changes they noticed in each of the girls. They were growing into adults. They were no longer kids. It reminded Maggie and me of our experiences during our first year at A&M.

One evening we got a call from Molly. The coach regretfully told her that she did not make the team. He then asked if she would like to attend the practices because circumstances sometimes change and even though it was unlikely, an opening could happen before the season got started. She wanted to know what she should do.

I asked her what she thought Grandpa would tell her to do. She said, "I think he would tell me to follow my heart." I said, "What does your heart tell you?" She told me that it told her if you don't take the chance, even if it is small, then you will never get the opportunity again. "Does that mean you are going to give it a try?" "Dad, I do not think Grandpa would want me to give up so I am going to do this for both of us."

My little girl was wise beyond her years. I was proud of her no matter how the spring training turned out. Molly was not only doing well in her studies, but she had also been working part time at a local veterinarian business. I told her she did not need to work. We had plenty of money for anything she needed. "Dad, I am not working there for the money. I am working for the experience." I was unable to argue with her reasoning.

Rex had us put aside and entire week where we reviewed every aspect of each business. We had charts, lists, bank statements, and projections. We reviewed the list of profit sharing accounts and oil price projections. It was five long days with vast amounts of information to go over. It was very productive. Rex was happy with everything. He felt the market for the rest of the decade would continue to have rising oil costs.

Even his ideas could not prepare us for the escalation in oil prices by 1980. Oil rose to over $35 per barrel before it started a slow decline to $15 by 1985. During that brief time, our companies made a fortune. Rex was presented with several buy-out offers. He even looked at a couple of them. In the end we kept our companies.

Molly kept us informed about baseball and her business classes. Mandy was enjoying her experiences with her boss at the clinic. One Saturday morning Molly called. She told us that the coach had called her in late on Friday and told her that he was impressed that she would come to practice and work hard even though she had no place on the team. He went on to say that she was a well-rounded player who could play more than one position and was a pretty good hitter. "I have been looking over our final picks for the season. The other coaches and I have made the decision to offer you a place on the team. Are you willing to come on board knowing that you will not be a starter?"

Maggie told him that she would be honored and would give a 100% for the team. She said he smiled and said, "I never doubted you for a minute." Maggie was beyond happy. “Dad, do you think Grandpa would be proud of me even though I am not a starter?” I told her that he would be proud of her for just trying. Maggie told her we were so proud to have a daughter like her.

A couple of weeks before the school year ended, Mandy called to say that the clinic had offered her a summer job. I told her that we would all be disappointed to not have her home for the summer. After talking a little longer, she had forgotten that this was Rex’s big birthday year and he would never get over it if she was not there. That closed the deal. She would be coming home.

Each of the twins offered their time to help Abby prepare for Rex’s big day. Maggie and Adam were trying to get last minute plans for the fall circuit dates. All our horses were ready and there were opinions circulating that Walker Ranch might take every prize in the fall. Our horses were just that good.

In the next three weeks, Usdi, Flower, and Bonnet all had their babies. We had two new colts and one new filly. As we had always done, we let the girls name each one. It was decided that Flower’s colt would be called Midnight. Usdi’s colt would be called River. And the little filly would be called Scarlet.

Maggie and Adam talked with Rex and told him that the training schedule was taking a lot of time, and they thought that we might consider selling a couple of the colts. Rex told them that it was their decision. Maggie said that Sunrise was not coming along as fast as they had hoped and it might be time to let him go as well.

Rex was allowing us to make plans and carry them out. He seemed happy with his life and was saying that he was still considering flying lessons. None of us took it seriously. Afterall, he was about to turn 70 years old.

Adam and Maggie started putting together a formal offering for selling some of our horses. The twins were helping Abby, and I was working with Andy on the daily business of our companies.

The big day came for Rex. Abby had planned a huge barbeque for all our employees. She had chartered a jet to bring people down from Afton. All our oil field crew members were there. A few important people from Dallas and San Angelo were invited. The girls had helped her plan activities for the children. The beginning of the summer heat was still a couple of weeks away, so we had some milder weather.

Before we started to eat, there were several people who spoke and thanked Rex for his impact on people and their lives. We had the Mayor of San Angelo, a gentleman from one of the Charities Rex supported, the directors from each of our two companies in Afton, and Eddie Most.

Members of the family thought we would save our comments for a private party we were having for Rex the next day. When the speakers were finished Rex thanked everyone for coming. Then he said, "I do not know about you, but I am hungry. Let's eat." He asked me to pray before we started.

It was a wonderful thing to see this many folks all gathered to celebrate the birthday of this man who had made all our lives better. At the private party the next evening Rex thanked Abby and the girls for putting together all the details of the party the day before.

Abby had told each of us in advance to prepare something personal to say to Rex instead of bringing a gift. One by one we shared our thoughts about this man that we all loved. There were some tears as people spoke. Rex had to wipe his own eyes more than once during the evening.

He deserved every nice thing said about him that evening.

The last person to speak was Abby. She thanked him for marrying her and sharing his life with her. She told him what a wonderful adventure it had been over these last 50 years. "God never allowed me to have children of my own, but you allowed me to have Maggie and Tommy and now Molly and Mandy. There could not have been a greater gift. Happy Birthday my Love."

That was the final straw, Rex could not hold the tears back. He walked over, held Abby tight and just sobbed. Rex Walker was a real man. His heart was forever giving, and he asked so little in return. I was so blessed to have had Rex, Grandpa and Andy as the men in my life.

It was beginning to weigh on me that they would not always be here. I wondered what my children would say about me someday in the future.

## CHAPTER THIRTY -EIGHT

It took Rex a few days to recover from his birthday party. It was an emotional time for him. A week later he called a meeting with Andy and me. When we were settled into the office chairs, he thanked us for the kind words at his party. Then he got down to business.

He felt that oil prices would continue to rise for a bit longer, but there would come a time when they would settle down and retract to prices that were more reasonable. He wanted us to consider several possibilities. The first was would we want to set and exit price where if oil prices hit a certain high, we would cut and run by selling our leases to a third party. The second was, would we take advantage of a high price and sell our wells. The third was would we keep ownership of the wells but sell 50% of rights to the oil to another oil company.

We talked for hours about the pros and cons of each plan. Rex said that we did not need to do anything and we could just keep it the way things are now. He was very convinced that the high price of oil would bring very strong offers for our wells. These offers would go down if oil prices started falling.

We agreed to discuss it more and see if we could put some guidelines or triggers in place that would allow us to implement an exit strategy once we had decided on a direction to go. The following months, as Rex had predicted, brought a continual rise in oil prices. The profits were beyond our dreams. We continued to drill and we were hitting good wells about 60% of the time. That corner of the ranch was sitting on very substantial oil-bearing layers.

Rex continued to pay geologists who produced the maps showing more potential areas to drill. Eddie had trained one of his very dependable men to take his place when he retired. His name was Noah Marsh. He had been making the decisions under the watchful eye of Eddie. He was doing very well and Eddie felt he was leaving us in good hands.

In December of 1977, Eddie Most retired. We had a huge party for him where Rex thanked him for all the years of faithful service. He praised Eddie for coming up with the idea of an all-Cherokee crew. Eddie was given a chance to speak. He thanked Rex for giving him the chance to be a part of a wonderful company. "Few white men have ever trusted and Indian to help run their company. I will never forget Rex Walker. He is the best man I ever met."

When Eddie finished. Rex said, "We just have one more thing to do. He handed Eddie a check for $1,455,000. It was the balance of his profit-sharing account. When Eddie looked at the check amount, he was unable to speak. He just hugged Rex and then walked away. There were lots of parting words from his crew and from the people who had grown to respect Eddie Most. It was the end of era.

The girls returned to school in the fall of 1978. Time was passing so fast that it was sometimes hard to believe. It was over twenty years since we had spent our time at A&M.

Molly played on the baseball team. She did not get to play in very many games. She remained a faithful player and team supporter. That would pay dividends in the coming seasons. Mandy was still working part time for the vet clinic. She loved her work. It appeared that she was heading that direction with her life.

Oil was still rising. It was approaching $25 per barrel. That was almost eight times what we were getting during the first years of our company. With the rise in oil value there were some costs that were increasing. A few were labor, drilling equipment, trucks, fuel, and pretty much anything that was considered necessary in the oil business. Even with increased costs we continued to have massive profits. The offers to buy our leases and wells continued to roll in. We had more discussion on an exit strategy but had not decided when we might act on any plans.

Adam and Maggie had completed a plan to sell several of our horses. There was no shortage of interested buyers. That year we sold Roamer, Midnight, Scarlet and River. With the girls away at school, we were entering our horses in fewer competitions. Our past successes still attracted buyers for any horses we were willing to sell.

1979 brought the highest oil prices in history. The price was approaching $30 per barrel. The new wells we completed were able to pay for themselves in just a few weeks. It was unbelievable.

When Rex bought his first company jet, he had a paved runway constructed. With some of our excessive profits he had several upgrades completed to our little airport. He kept the first little jet and shortly after the improvements were made to the airstrip, he purchased a larger one that could hold more passengers and had the capability of flying greater distances without needing to refuel.

Molly called one evening and was so excited. Her coach had said that she would be playing an entire game the coming Friday. We told her how proud we were and wished her good luck at the game. When we told Rex, he told Abby to call the school and get all the info. "We are taking the new jet, and we are going to go see that game."

Molly was moved to tears when she saw us before the game started. Rex told her that no power on earth could have prevented us from being there to cheer for her. Mandy knew we were coming and had arranged for our tickets and seating. We had a wonderful time. The team lost the game

However, Molly was able to get one hit and caught two fly balls in center field where she was playing.

We took her out for dinner after the game. As we were leaving for the night, Molly grabbed Rex around the neck, kissed his cheek and said, "Thank you Grandpa Rex. This is the best day of my life". Rex tried to hide his emotion, but it was difficult. He loved those girls as his own.

The next morning, we all went with Mandy to see the vet clinic where she was working. It was a very modern and well-equipped facility. Mandy showed us around the entire building and introduced us to the other staff members. The main veterinarian, Dr Whitman had nothing but praise for Mandy. We spent a couple of hours with her before we headed back to the plane.

On the flight home Abby looked at Rex and said, "That was a good thing you did for the girls." He just nodded. None of us could imagine life without Rex. Somehow, I think he felt the same about each of us.

In the fall of 1980, the girls returned for their junior year of school. Oil hit $35 per barrel, and we had an additional seven good wells go online. Rex was really wanting us to decide if we were going to sell some of our assets or just ride this wave and see where it went.

We spent weeks going over figures and projections. As we narrowed down our thoughts, we came up with a plan where we could sell some of our wells that were produced during our first years of drilling. The production was still level, and a prospective buyer would not think we were selling wells that might have a short lifespan.

Andy asked Rex what we would do with the money if we sold the wells. His answer showed that he had given it quite a bit of thought. He said, "I would like to set up a scholarship fund for the children of the ranch personnel and the businesses in Afton. Their schooling would be free, providing they kept up their grades. However, they must attend A&M. Secondly, I want to purchase more land and set up a wildlife preserve. And lastly, I want to build a large Veterinary hospital in San Angelo. I want it to be the best facility in this part of the U.S. I want it to be the home of Dr Mandy Meadors.

When I told Maggie that evening, she began to cry. No one individual had cared so much for the members of our family. Rex was leaving a legacy made up of those he loved. It was a wonderful plan. We made a pact to not mention this to anyone. It was a plan Rex made, and he should be the one to tell others.

We spent a couple of months looking at which wells we might sell and what ranch land we would want to buy with the sale money. We studied the lands bordering the existing Walker Ranch property. We flew over them in one of our planes. We created lots of aerial photos that we could study. Natural water sources was one of our prime concerns.

None of these decisions were easy. Each one had its own set of challenges. In the end, we put together a group of 25 wells. We decided to let fate guide us from there. The price we placed on the sale was very high. Even with $35 a barrel oil, it would take some time to ever recover enough oil to re-coop the purchase price. If there were no bidders, we were going to keep the wells.

We had multiple interested parties, who ended up bidding against each other. In the end, we sold the wells for much more than we had originally asked.

We had made offers on the ranches we were interested in purchasing. Three of the four we wanted agreed to sell at the prices we had offered. The total land we purchased added more than 34,000 acres to the existing ranch. They included more than 15,000 acres of good grazing land, 14 lakes, and several hay barns. Rex was very pleased.

The three of us spent the next few weeks looking for a site in San Angelo where we could build the new Vet hospital and school. The city owned some undeveloped land. When they heard why we wanted the land, they made us an offer that was too generous to pass up. The deal was struck and Rex started looking for an engineer to draw the plans. He held off on the final plans as he wanted to include input from Mandy when she came home for the summer.

The twins were turning twenty this year. Abby wanted to have a big celebration. Everyone agreed that she could plan the party. July was going to mark a milestone. We no longer had teenagers. We got another surprise when the girls came home. Molly was getting cozy with one of the baseball players at A&M. He was not from Texas. His family owned several pecan orchards in Georgia.

She asked if he could come to the birthday celebration. We let Abby make that decision and of course it was yes. It appeared that we were going to meet Brad Williams at the birthday party.

Rex planned on telling Mandy about the new hospital and school. He was going to take her to town and just the two of them would spend the day together. He was more excited about that trip than he had been about selling some of our wells.

Andy was helping Rex with setting up the wildlife preserve. I was creating a plan on how we would be farming the acres we wanted to develop. It was shaping up that we would concentrate on cotton, wheat, and alfalfa. The lawyers were creating the scholarship funding program. We had made good progress in a short period of time.

Mandy spent most of her days working with the ranch vets and doctoring livestock. Rex asked her to take a day off and ride to San Angelo with him. They were gone until the evening. When they returned, Mandy could not wait to tell everyone what had happened. She would not even let us finish our dinner.

"Well do you want to hear about my trip with Grandpa Rex?" It did not matter what we said, we were going to hear all about it. "He took me to lunch and then to this property in town. It had a sign on it that read "future home of Walker Veterinary Hospital and School". Then he told me that if I got my master's degree, I would be one of the primary doctors and instructors." It was hard for her to control her excitement.

Now all three parts of the plan Rex wanted were falling into place. There was one last thing for him to do. He wanted to help Molly with her future. Neither of them knew exactly what shape that future would take. For now, he was willing to wait until Molly figured out where she wanted to go and what she wanted to do with her life. Rex hoped it would be something close to the ranch if it was not on it.

The day before the big birthday party, Brad Williams arrived on the ranch. Molly met him at the headquarters and then she brought him up to our house where we met him for the first time.

He was tall and lanky. Very tanned with brown hair and eyes. He was wearing a baseball cap that he removed when Molly introduced him to Maggie and me. He looked like most of our ranch hands. Molly did not treat him like a ranch hand. It was easy to see that there were emotions on both sides. We spent the afternoon talking with him and getting to know a little of his history. Rex wanted us to bring him to dinner that evening, so we had a little light lunch.

He asked a few questions about the ranch, and I asked about their orchards. It turned out that they had 700 acres of pecans. His grandfather had started their ranch, and his father was now running the operation. He told us that he was expected to graduate and then return to help run the family operation. When he said that, I looked at Molly to see her reaction. There was none so it must have meant that she knew his plan.

At dinner Rex asked Brad what kind of soil would make a good pecan orchard. The two of them talked for a while. I could see the gears in Rex's head turning. The evening was very pleasant. Everyone liked Brad. Mandy was very supportive of her sister's choice in a guy.

The big party went off without any problems. Abby had done a wonderful job of planning every detail. It was not as big as the party we had for Rex on his $70^{th}$, but it was still wonderful. That night after everyone went their own way and Maggie and when I were alone, I asked her if it was possible that our girls were twenty years old.

Brad stayed for a couple of days before heading home. Molly had him riding horses both days as she showed him around the ranch. From what she told us later he was very impressed with the size and scope of the operation.

After Rex had talked with Brad, you were able to see that he had more than a casual interest in pecans. For the next few weeks, he had me researching everything that I could find about pecan orchards. There was a huge pecan growers convention coming up in the fall and Rex had me make reservations for us to attend.

# CHAPTER THIRTYY-NINE

The girls headed off to their last year of school in the fall. Brad came home with Molly and Mandy at Christmas. Maggie and I were not surprised. One evening after dinner Brad asked if he could talk to me alone.

While we sat on the front porch, he began to tell me, how much he loved Molly. "Tom, I want to marry Molly when we graduate. It would mean more to me than you realize, but I would like to have your blessing. I have already told my parents. We visited them one weekend and they loved Molly. My dad wants us to come back to the ranch after we are married and help him farm our orchards."

"Well Brad, Molly is a grown woman, and she can make her own decisions. I trust her judgment. The part about going to help your father is a bit of a snag. You see Rex has developed a great interest in pecans and he wants you to help him develop a couple of thousand acres of land we recently purchased. Molly has a home here. It was where her grandparents lived. It will be hers when she marries."

"I do not want to spoil your dad's plans, but it would be very important to us to have Molly living on the ranch. I would be happy to talk with your father about our plans for pecans. This is an opportunity for you to start a pecan business from scratch. We have the land, finances, and manpower. If it works well, we can easily double the number of acres. You can build a legacy here or go back and work on land that your father has already developed."

Brad looked shocked. "I do not know what to say. Does Molly know about this?" I told him that she did not. "I do not want to give you an answer until we have had a chance to talk. Would that be ok with you?" I told him that he was being wise and that If he and Molly decided to move forward with our plans for orchards, he should invite his mom and dad to the ranch so we could all talk together.

He liked my thoughts and within a couple of days, he and Molly met with Rex and Abby. Rex explained that we had attended the Pecan Growers convention and that we were not charging into this business with foolish goals. He explained about our purchase of the 15,000 acres and our desire to do something new with that land. I do not think Brad was prepared to be overwhelmed by Molly's Grandpa Rex. He sat there and for a long time did not even have the words to express his feelings.

Rex said, "Before you make the final decision, I would like to invite your mom and dad up to the ranch. I would like your dad's input, and I want them to feel part of this project." Brad seemed to feel some relief to know that Rex wanted to include his family. Rex told him to contact his parents, and he would send a plane down to get them. That way we could all talk about this before they needed to return to school.

That is exactly what happened. Brads parents, Robert and Ann arrived at the ranch two days later. Rex had Brad and Molly show them around the ranch for an entire day. That evening we all met for dinner at Rex's home. Abby had it planned perfectly. It was a wonderful meal, and we got to know Robert and Anne a little better.

After dinner, the women stayed together and the men went outside and sat on the porch. Rex explained to Robert that he had a lot of land that needed to be developed. He wanted to give Brad the chance to be in on the ground floor. Robert listened and finally said, "Well Rex, it is not often that a young man is given the opportunity to be a part of what you are trying to do. It would be selfish of me to hold him back. I will give him my blessing if this is what he wants to do."

Rex smiled and said, "Robert I am a cattle rancher and an oil man. We try to stay up with the changes and be on the cutting edge of modern practices. We do some farming on a few thousand acres. This orchard thing will be new to us. What would you think about being part of the operation?"

Robert did not even know what to say. Rex continued, "We will need the experience that you have accumulated with your orchards. I would like to offer you a percentage interest in our new business. I do not want you to run it. That will be Brad's job. I want you to be our consultant and guide as we move into this new business. What do you think?"

Robert had to think for a moment. "Well Rex everything that I have heard about you is true. You do not mess around. When you go for something, it appears to be something on a large scale. To offer my son this opportunity is more than I could have wished for him. To include me is more than expected. I will talk to my wife tonight. But I can tell you that I would be a fool to turn this down. Unless she has some huge reason that we should not do this, you can count our whole family in on your plan."

In true Rex fashion the whole thing was planned in a couple of days. The wedding, moving into Grandpa's house, Robert helping from ground zero, and Brad becoming a member of the ranch family businesses.

We all flew down for the graduation. It was a milestone in all our lives. The two kids Rex had helped through college were now his family and their two daughters were grown up and adding another generation to the Rex and Abby legacy.

In the last few months before the kids graduated, Robert had been supervising the preparing of the land where we would be planting trees in the winter. Rex was very happy with the progress.

The wedding took place in 1981 at Rex and Abby's home just as ours had. Maggie cried through most of the ceremony as did Abby, Anne and Stephanie. They had a weeklong honeymoon in the new house that had been redecorated for them by Abby and Stephanie. Brad was anxious to get anxious to get started on the orchards. Over the next few months, he was able to work with his dad and before the trees arrived, we had all the land prepared and our irrigation in place. Rex was more than happy at the progress.

With the orchards well on the way to happening, Rex concentrated his time with Mandy on planning the final touches to his plan for the teaching veterinary hospital. Maddy was in her glory working with Grandpa Rex on something that was so near to her heart.

When she and Rex had their wishes committed to paper, the architects were given the green light to finalize the plans and submit them for approval so that the building process could go forward. Rex took advantage of his free time and spent most of it with Mandy. Her love of animals gave them a common goal of getting the hospital up and running. He must have been feeling his age because his drive to see the hospital and orchards going strong, was his priority.

The man had built an empire of businesses. His legacy was to see those he loved, continue running the dreams that he had built one by one. By the spring of 1982, the first one thousand acres of pecans had been planted, the hospital had a grand opening and Rex had his 75th birthday.

Adam had continued training our horses. We were no longer competing on a national level. There was still a huge demand for any horses that we offered for sale. Our oil drilling was continuing to produce winners about sixty percent of the time. There had been some retreat in the prices, but we were making huge profits even at the current rate of $28 per barrel. Rex had been very wise to sell some of our wells at what turned out to be the peak of the recent markets.

In 1983, Brad planted another thousand acres of pecans. Mandy and the Walker Veterinary Hospital and School were both becoming well known throughout Texas. Andy and I had our hands full looking after so many of Rex's projects. Brads father Robert had been keeping a close watch on the orchards. He and Brad worked very well together.

In 1987 we had some major things happen on the ranch. Rex turned 80 years old. One of the largest meat companies in the world made an offer to purchase all our assets in Afton. Mandy had hired a new doctor who she was now dating.

We had our first harvest of pecans. We only harvested the first one thousand acres. The second thousand were still a bit young. The oil prices had bottomed out at $15 per barrel in 1985 and they were on the rise again.

Rex made a deal for the companies in Afton with the provision that the new owners must purchase their livestock from Walker Ranch. The biggest thing in our lives that year was the announcement that Molly and Brad were going to be parents in a little over six months.

On March 10, 1986, little Tommy Rex Williams was born. That poor kid did not have a chance. There were two grandmothers and one great grandmother who were fussing over who was going to hold him next. The grandfathers did not have a chance. We were all happy and proud of the new addition to our family.

In 1988 we harvested the total of two thousand acres of pecans. It was wonderful to see how all those years of labor had produced some of the most beautiful pecan orchards in Texas. Brad was turning into a wonderful farmer. We were all proud of him. Because of our grower size, he had been elected as one of the officers in the Texas Pecan Growers Association.

Mandy and her doctor boyfriend had announced that they were getting married. Molly, not to be out done by her sister announced that she was having another baby. The new baby, Jamie Roy Williams was born on January 15, 1989. Mandy and Dr. John Forman were married on May 12$^{th}$ in Rex and Abby's home. Despite being 81 years old, Abby wanted to help plan everything so the wedding would be perfect.

Rex purchased them a new home in town for a wedding present. He would have much rather had them living on the ranch. It was not practical considering the commute.

In May of 1990 we faced a crisis. Rex was riding one of his favorite horses and for some unknown reason was thrown off and landed on some rocks near a creek where he had been riding. He broke a hip and ended up in the hospital. During his time in the hospital, he developed an infection in his blood and what should have been a few weeks of recovery turned into a long process. Mandy checked on him every day. Although he was weak, he seemed in good spirits.

The doctors confided in Mandy that they were having a very hard time getting control of the infection. Mandy did some checking and it was decided that it would be best to transfer him to Dallas where specialists in this area had a practice. We sent one of our planes to take him to Dallas. Mandy and another nurse rode along on the flight.

At first his condition improved. After a few days at the hospital in Dallas, the specialist spoke with Mandy and told her that it was getting very serious and that if they were unable to get control, some of his organs would start shutting down. Mandy relayed the information to us and Abby was flown to Dallas to be with Rex. Stephanie stayed with her every minute.

After not getting any better, Rex asked if we could all come and see him. It was a sad flight. No one said anything. We were each lost in our own thoughts. There must be something that could be done. This man we loved could not just go away. How could being thrown from a horse result in this situation?

The doctor told us that Rex was very weak and that we needed to limit our visit with him. Each family got a few minutes. When they came out, every single person was crying.

When Maggie and I got our turn with the twins, he smiled as we entered his room. "I saved you for last because you have made my life complete. My heart is full of memories. I want you to know that I am not sure that I knew what love was all about until the time you became part of my life. Thank you for allowing me to be a father and grandfather and great grandpa. Whatever you think I may have done for you is far outweighed by what you have done for me. Roy taught me how to say Donadagohvi. That is the best saying...we will see each other again. God loved me enough to give me each of you. Wado."

He hugged and kissed us each goodbye. We never saw him again. He died in the early morning hours the next day. None of us was able to function for many days. It was over. There would never be another day that we would hear his voice or see that classic smile. It was the end of an era. Abby told us later that he wanted to be buried beside Grandpa and Grandma. We respected his wishes and had a little private time when we buried him at that special place.

There was a huge celebration of his life. People came from all over the country. Rex was respected and loved by so many people. There were several speakers at his service. Some from business and others from government. I think the ones that would have mattered the most to him were those employees who had served him for so long. Many of them cried as they spoke of what Rex had done for them. Abby ordered a gravestone. It read " Rex Walker, he loved with his whole heart." For a very long time, there was no joy in life. Rex was the glue that held everything together. He loved in such a way that each of us felt we were special to him. Abby tried to stay busy with little Tommy. He became her focus. Everyone pitched in to watch after Abby. She was still doing well but age was starting catch up to her.

Somehow life seemed to continue. The orchards had a bumper crop in 1991. Oil prices continued to climb. Mandy and John had a little girl they named Emma Lee. Andy turned Seventy-one. He was still active in all our businesses. Abby seemed to be empty at times. Rex had been her whole life. We never knew anything about their families. It had always been one of those things that was not discussed. We did not even know how long they had been married.

By the time Emma was able to walk and was talking, Abby seemed to have lost any purpose. She spent most of the time alone. If one of us went to check on her she was always glad to see us but there was always a cloud of loneliness that seemed to hang over her more each day.

When we lost Rex, the length he had gone to in making sure there were legal documents in place to protect each of us from potential problems, was amazing. Abby inherited everything. All the corporations and businesses which were held in the Walker Trust were managed by Andy and me. All our children were mentioned in the will and there were things put in place to give each of them a start in life no matter what they chose to do for a career.

Upon the death of Abby, the assets of trust in their entirety passed on to us. There were guidelines for certain charities that Rex wanted to give continued support. The funding for A&M scholarships would remain intact. The support for the Walker Veterinary Hospital and School would have annual funding that was perpetual.

For Andy and me, it became a fulltime effort to manage things the way Rex had done. Many times, in our meetings one of us would say "How do you think Rex would have wanted us to do this?"

Brad had revived our school baseball team, and we all had a great time watching them play. Mandy and John had given of themselves and their time to make the Hospital into a nationally respected institution. Over the years, they sought out promising young people to staff the halls of their business. Rex would have been proud of all their efforts.

In 1993, Maggie and I turned sixty years old. Andy and Stephanie turned eighty and we lost Abby at eighty-six. She was never herself after Rex died. I think she just kind of gave up on life. She had a written request to be buried beside Rex and that is what we did. Her gravestone said, "Abby Walker beloved mother, grandmother and great grandmother."

It was so different now. The two people who made a life, lived a dream, and created hope for so many, were gone. It was if the dreams had stopped when we lost them. Those of us left behind, were just managers of all they had created. We were not the dreamers.

For a long time after they were gone, their house sat

vacant. One day Andy told me it was not right to let it remain empty. He and Stephanie felt that Maggie and I should move in and make it a happy place again. He said, "that is what they would have wanted."

We moved in and each year on their birthdays, we had a party for Rex and Abby. Stephanie and Maggie put together a huge inventory of pictures that were taken over the years. They cataloged them by year. Each year we added new pictures to our memories.

In 1994 Rex was inducted into the Texas Cattlemen's Hall of Fame. It was a great moment to have so many ranchers recognize his contribution to the Texas cattle business. In the fall we lost Maggie's father Andy. It was a huge loss to me. He and I had worked on so much ranch business together. I had known him my entire life.

First Grandpa, then Rex and now Andy. The three men who had helped shape me. I was alone. For many weeks, I was unable to concentrate on my work. I would leave in the morning and just drive out to the special place. I would sit there and many times just cry and tell them that I did not know what to do now. Those tombstones had lots of tears on them. Maggie started to worry about me.

One night she said "Gule, there are lots of people who are depending on you to keep things going. Rex picked you to be in this exact place that you find yourself in today. Are you going to show him that he was wrong to put his faith in you? You are the new Rex. People are looking to you for direction and leadership. We have children and grandchildren to protect and influence. Rex, Grandpa and Andy believed in you. I believe in you. Grandma said she had an oak and an acorn. It is time for the acorn to become the oak."

We agreed from that day forward, she would never call

me Gule. My goal had become the role of the oak for my family. As the grandchildren grew older, they were taught about their grandparents and great grandparents. They knew who Roy and Ida were. They knew all about Grandpa Rex and Grandma Abby. Their roots were this ranch and the individuals who created something from very little.

In 1996 we lost Stephanie. She passed peacefully in her sleep. She wanted to be buried next to Andy. Her gravestone read, "Stephanie Scott, mother and grandmother who was loved by many." It was full circle. Maggie and I were all that was left of a former time. We knew the players and their roles in this thing called life.

Many evenings, Maggie and I would sit on the front porch where we had spent so much time with those we loved. We would look down at all the things that working hands had created over the decades. God had blessed us beyond our dreams.

On that porch, futures had been planned and lives were changed. And on that porch an acorn had become an oak. Donadagohvi.

www.ingramcontent.com/pod-product-compliance
Lightning Source LLC
LaVergne TN
LVHW010712061126
851857LV00014B/1149
* 9 7 9 8 2 3 4 2 0 3 5 2 6 *